Charles Barney Cory

How to Know the Shore Birds of North America

Charles Barney Cory

How to Know the Shore Birds of North America

ISBN/EAN: 9783337217853

Printed in Europe, USA, Canada, Australia, Japan

Cover: Foto ©berggeist007 / pixelio.de

More available books at **www.hansebooks.com**

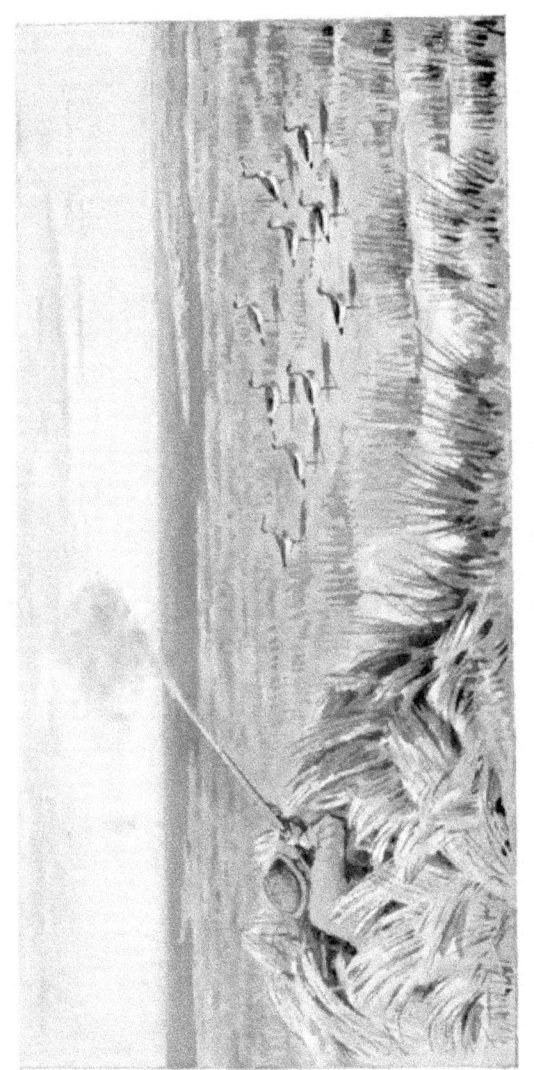

SHOOTING YELLOW-LEGS.

ALFRED MUDGE & SON, PRINTERS,
24 FRANKLIN STREET.

CONTENTS.

	PAGE
PREFACE	5
HOW TO MEASURE A BIRD	7
GLOSSARY	9
INDEX TO KEY	11
KEY TO THE SHORE BIRDS OF NORTH AMERICA KNOWN TO OCCUR SOUTH OF GREENLAND AND ALASKA	13
GROUP 1. Birds having wings measuring from 3.25 to 3.75 inches long	13
" 2. " " " " " 3.75 to 4.50 " "	14
" 3. " " " " " 4.50 to 5.50 " "	16
" 4. " " " " " 5.50 to 6.75 " "	21
" 5. " " " " " 6.75 to 9 " "	24
" 6. " " " " " 9 to 12 " "	28
FAMILY PHALAROPODIDÆ. The Phalaropes	31
" RECURVIROSTRIDÆ. The Avocets and Stilts	35
" SCOLOPACIDÆ. The Snipes, Sandpipers, etc.	38
" CHARADRIIDÆ. The Plovers	71
" APHRIZIDÆ. The Surf Birds and Turnstones	79
" HÆMATOPODIDÆ. The Oyster Catchers	82
" JACANIDÆ. The Jacanas	85
INDEX	87

PREFACE.

The present work is intended to meet the wants of a large number of persons, especially sportsmen, who are interested in birds and would like to know their names, but often find it no easy task to identify them by the "bird books." To all such I offer this Key, in which the species are arranged in groups according to size, and believe it will enable any one unfamiliar with birds to identify with comparative ease any species of our North American Shore Birds.

Birds vary so much in size that the *length* of any one specimen cannot be accepted as a standard for others of the same species. The length measure is, nevertheless, of value to enable us to form an approximate idea of the size of the bird. The length of the *wing* is, however, much less variable, and is an important aid to the identification of many species. In fact, the variation is so small and constant that, allowing for possible extremes, they may be arranged in groups according to length of wing. The identification of any species then becomes a very simple matter, as usually the birds contained in each group are so few in number that characteristic differences in each species are easily indicated.

Let us assume, for example, we have before us a bird which we wish to identify. We first measure the wing (see directions for measurements, illustrated, page 7). We find the wing measures nine inches long. We now turn to the "Index to Key" (page 12), and find that Group 6 contains birds having the wing measuring from nine to twelve inches long, and is divided into three sections; Sections 1 and 2 comprising birds having four toes, and Section 3, birds with three toes. Our bird has four toes, so we look for it under Section 1 or 2. We find that birds under Section 1 have the bill curved downward, and birds under Section 2 have the bill curved upward, or nearly straight. Our bird has the bill curved downward. We therefore look for it under Section 1. We find but two birds included under Section 1; one with axillars reddish brown, marked with black, and the other having the axillars banded with slaty brown and dull white. As the bird before us has slaty brown axillars banded with dull white, it must be the Hudsonian Curlew, *Numenius hudsonicus*. The axillary plumes are often a very important aid in determining a species, and the beginner should learn where to look for them and to recognize them at a glance. (See cut, page 10.)

All measurements of birds are given in inches and fractions of an inch. The following diagrams will illustrate how a bird should be measured, and the chart (page 10) will be useful to the young student of ornithology who may not be familiar with the technical terms used in describing birds. Such terms as primaries and axillars should be learned at once. It is customary to indicate the sexes by the signs of Mars and Venus; the male, of course, being given that of Mars, ♂, and the female, Venus, ♀.

In preparing the Key, a large number of birds were examined and measured. In this connection, my thanks are due to Dr. J. A. Allen, of the American Museum of Natural History; and to Mr. Robert Ridgway, of the Smithsonian Institute, for the loan of many specimens for examination; and especially to Mr. William Brewster, for free access to his magnificent collection in Cambridge.

The illustrations are the work of Mr. Edward Knobel.

CHARLES B. CORY.

BOSTON, MASS., June 20, 1897.

HOW TO MEASURE A BIRD.

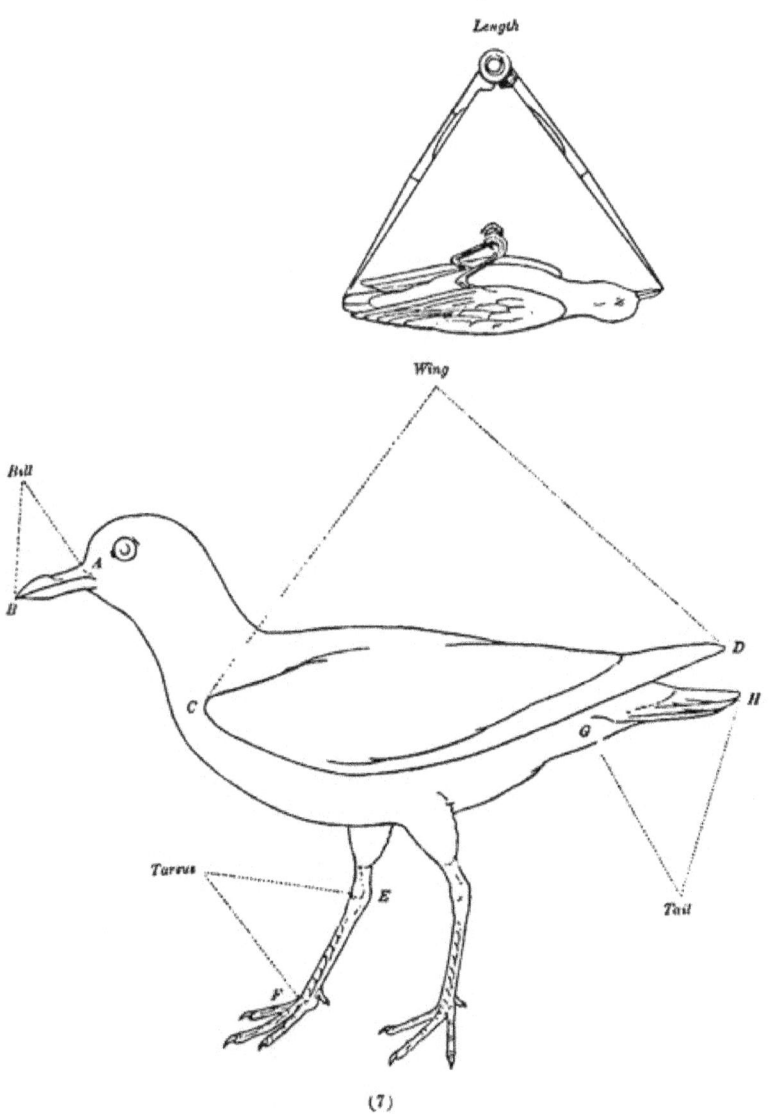

Wing. Distance from carpal joint C (bend of wing) to the tip of the longest primary D. See cut on page 7.

Length. Distance in a straight line from the end of the bill to the tip of the longest tail feather. (Occasionally the middle feathers are much elongated, as in the Old Squaw and Pintail Duck, and in other families of birds, such as Phæthon and Stercorarius. In such cases it is well to give the length from bill to longest tail feather, and also to end of outer tail feather.)

Tail. Distance from the tip of the longest tail feather to its base (the point where it enters the body).

Bill. The distance in a straight line from where the bill (upper mandible) joins the skin of the forehead (A) to the tip (B). (There are a few exceptions to this rule, such as birds with frontal plate, etc. Some curved bills are measured along the curve of the culmen, and at times it is advisable to measure from the nostril to the tip of the bill, but in such cases it should always be so stated.)

Tarsus. Distance in *front* of the leg from what *appears* to be the knee joint (end of tibia) to the root of the middle toe. All measurements are given in inches and fractions of an inch.

GLOSSARY.

Nearly all the terms used in describing a bird may be more easily and clearly understood by examining the accompanying figure than from a written description; a few, however, may require a word of explanation.

Mandibles. — Some authors use the word *maxilla* for the upper half of the bill, and *mandible* for the lower. I prefer, however, to describe the two halves of the bill as upper and lower mandible.

Culmen. — The ridge of the upper mandible.

Gonys. — Lower outline (middle) of under mandible.

Unguis. — The nail on the end of the upper mandible; very pronounced in several families of water birds, — Ducks, Pelicans, and Petrels.

Axillars or Axillary Plumes. — Several elongated feathers at the junction of the wing and body. (Lat. axilla, the arm-pit.)

Speculum. — A wing band or patch (usually of a different color from the rest of the wing), formed by the terminal portion of the secondaries; very noticeable in the Ducks.

Tarsus. — Extends from the root of the toes to the end of the tibia (what *appears* to be the bend of the leg or knee; but which is, in reality, the heel joint).

Superciliary Stripe. — Stripe over the eye.

GLOSSARY.

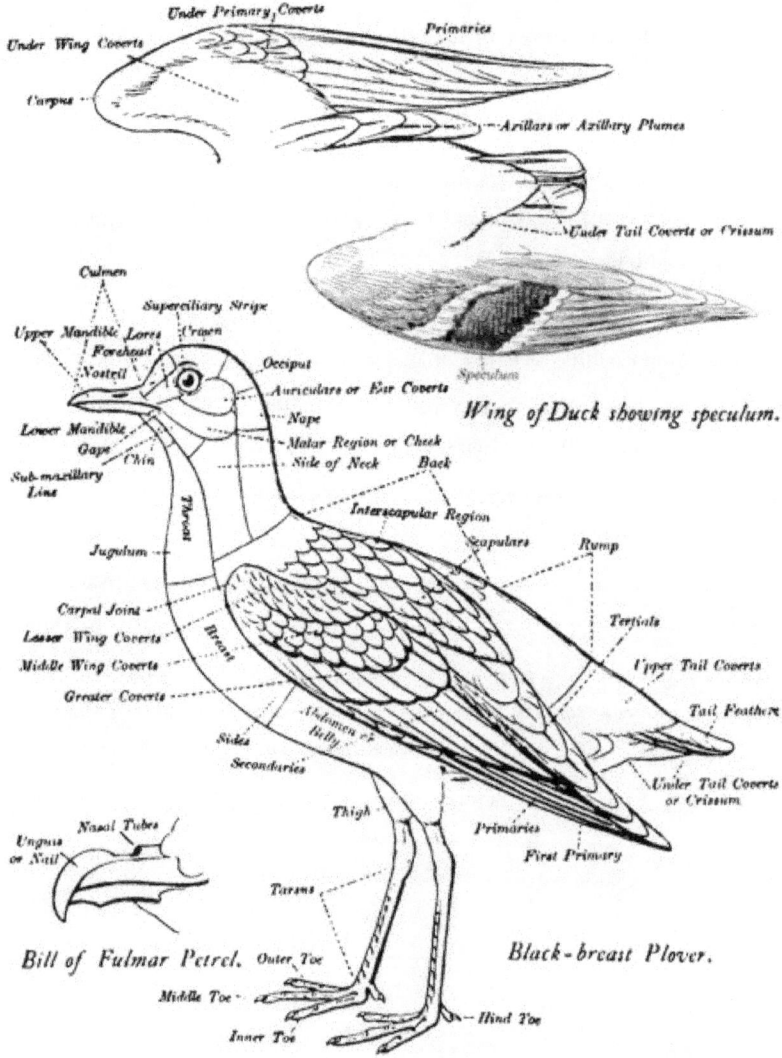

INDEX TO KEY.

		PAGE
Group 1.	Birds having wings from 3.25 to 3.75 inches long	13
Group 2.	Birds having wings from 3.75 to 4.50 inches long	14
	Section 1. Toes 4, with small lobate webs	14
	Section 2. " 4, " " webs (not lobate) between toes at base . . .	14
	Section 3. Toes 4, without webs	14
	Section 4. " 3, bill under .60	15
	Section 5. " 3, " over .60	15
Group 3.	Birds having wings from 4.50 to 5.50 inches long .	16
	Section 1. Toes 4, a small web between toes; bill under 1.75 .	16
	Section 2. " 4, " " " " " over 1.75	17
	Section 3. " 4, without webs; bill over 2 . .	17
	Section 4. " 4, " " " over 1.10 and less than 1.90	18
	Section 5. Toes 4, without webs; bill under 1.10	19
	Section 6. " 3, bill over .60	20
	Section 7. " 3, " under .60	20
Group 4.	Birds having wings from 5.50 to 6.75 inches long . .	21
	Section 1. Toes 4, a small web between outer and middle toe; bill over 1.80	21
	Section 2. Toes 4, a small web between outer and middle toe; bill under 1.80 . . .	21
	Section 3. Toes 4, with small lobate webs	22
	Section 4. " 4, without webs; bill over 2 .	22
	Section 5. " 4, " " " under 2 .	22

(11)

INDEX TO KEY.

		PAGE
Group 5.	Birds having wings from 6.75 to 9 inches long .	24
Section 1.	Toes 4, with more or less web; bill curved upward or straight, — bill over 2.60 inches long . . .	24
Section 2.	Toes 4, without webs; bill nearly straight, — bill over 2.60 inches long	25
Section 3.	Toes 4, with small web; bill curved upward or straight, — bill under 2.60 and over 1.50 inches .	25
Section 4.	Toes 4, with small web; bill nearly straight, — bill under 1.50 inches	26
Section 5.	Toes 4, with small web; bill curved downward, — bill over 2 inches	27
Section 6.	Toes 4, without webs; bill under 1.50	27
Section 7.	Toes 3	27
Group 6.	Birds having wings from 9 to 12 inches long	28
Section 1.	Toes 4; bill curved downward . .	28
Section 2.	" 4; " " upward or nearly straight .	29
Section 3.	" 3; " nearly straight	29

KEY TO THE SPECIES.

GROUP I.

Birds having wings from 3.25 inches to 3.75 inches long.*

Tringa minutilla.

No web between toes.

Least Sandpiper.
Tringa minutilla.
See page 48.

Toes with small web; bill, usually under .85.

Semipalmated Sandpiper.
Ereunetes pusillus.
See page 51.

Toes with small web; bill, usually over .85.

Ereunetes pusillus.
(Foot.)

Western Sandpiper.
Ereunetes occidentalis.
See page 52.

* For directions for measurement, see page 8.

GROUP II.

Birds having wings from 3.75 to 4.50 inches long.

Section 1. Toes, 4, with small lobate webs.

Phalaropus lobatus.

More or less rufous on sides of neck.

Northern Phalarope.
Phalaropus lobatus.
See page 33.

Section 2. Toes, 4, with small web (not lobate) at base.

Ereunetes pusillus.

Bill, under .85: back, not greenish olive; bill, entirely black.

Semipalmated Sandpiper.
Ereunetes pusillus.
See page 51.

Bill, over .85; back, not greenish olive; bill, black; no white patch on inner webb of third primary.

Western Sandpiper.
Ereunetes occidentalis.
See page 52.

Small web between outer and middle toe; bill, over .85; **back, greenish olive,** sometimes banded; under mandible, pale yellow (in life); third primary and inner primaries with patch of white on inner web.

Spotted Sandpiper.
Actitis macularia.
See page 66.

Section 3. Toes, 4, without web.

Tringa minutilla.

Belly, white; bill, black.

Least Sandpiper.
Tringa minutilla.
See page 48.

Section 4. Toes, 3; bill, under .60.

Aegialitis semipalmata.

Bill, orange at base, the tip, black; legs, dull flesh color; a black stripe from bill passing under eye.

Semipalmated Plover.
Aegialitis semipalmata.
See page 75.

Bill, orange at base, the tip, black; the legs, orange yellow; two middle tail feathers, tipped with white; no black stripe from bill to eye; black breast band not confluent. **Species not found west of the Rocky Mountains.**

Piping Plover.
Aegialitis meloda.
See page 75.

Bill, orange at base, the tip, black; legs, orange yellow; middle tail feathers, tipped with white; no black stripe from bill to eye; **a continuous black band on breast. Species not found west of Rocky Mountains.**

Belted Piping Plover.
Aegialitis meloda, circumcincta.
See page 76.

Aegialitis nivosa.

Bill, entirely black; legs, slate color; two outer tail feathers, entirely white; two middle feathers, **not** tipped with white; no black stripe from bill to eye. Ranges from Texas and Kansas west to the Pacific Ocean; casual in Western Florida and Cuba; **not known to occur on the Atlantic Coast.**

Snowy Plover.
Aegialitis nivosa.
See page 76.

Section 5. Toes, 3; bill, over .60.

Aegialitis Wilsonia.

A very small web between outer and middle toes; bill, large and thick; a band of black (male) or brown (female) on breast.

Wilson's Plover.
Aegialitis wilsonia.
See page 77.

No web between toes: bill, not thick.

Sanderling Sandpiper.
Calidris arenaria.
See page 53.

Calidri- arenaria.

GROUP III.

Birds having wings measuring from 4.50 to 5.50 inches long

Section 1. Toes, 4, a small web between toes; bill, under 1.75.

Bill, under 1.75. Tarsus, over 1.30. All other species in this section have the tarsus less than 1.30.

Stilt Sandpiper.
Micropalama himantopus.
See page 42.

Axillars.
Totanus solitarius.

Tarsus, under 1.30. Back, dark olive spotted with white, or brownish gray spotted with dull white, according to season. **Axillars, heavily barred;** a small web between the outer and middle toe.

Solitary Sandpiper.
Totanus solitarius.
See page 59.

Back, greenish olive, *sometimes* barred with black; **axillars, white** without bars. At some seasons underparts with round black spots; a small web between the outer and middle toe.

Spotted Sandpiper.
Actitis macularia.
See page 66.

Back, heavily streaked with black and tawny; **belly, reddish brown,** showing more or less white; **toes, with small lobate webs.**

Red Phalarope.
Crymophilus fulicarius.
See page 32.

Crymophilus fulicarius.

Back, grayish, streaked with tawny; belly, white; toes, partly webbed; bill, under 1.05; tarsus, under 1.

Northern Phalarope.
Phalaropus lobatus.
See page 33.

Phalaropus
lobatus.

KEY TO THE SHORE BIRDS OF NORTH AMERICA.   17

Back, grayish, marked with **chestnut brown; belly, white; bill, over 1.05;** tarsus, over 1.

Wilson's Phalarope.
Phalaropus tricolor, female.
See page 34.

Phalaropus tricolor.

Back, grayish, mottled with dusky or whitish; bill, over 1.05; tarsus, over 1 inch.

Wilson's Phalarope.
Phalaropus tricolor, male.
See page 34.

Section 2. Toes, 4, a small web between outer and middle toes; bill, over 1.75.

Macrorhamphus griseus.   Macrorhamphus griseus.

Bill, over 1.75. Axillars, white, barred with dark brown; rump and tail, white, spotted and banded with black.

Dowitcher, Red-breasted Snipe.
Macrorhamphus griseus,
and
Long-billed Dowitcher.
Macrorhamphus scolopaceus.
See page 41.

Section 3. Toes, 4, without webs; bill, over 2 inches long.

Axillars, rufous brown, without bars; belly, buff color.

Woodcock.
Philohela minor.
See page 38.

Axillars, barred black and white; belly, white; upper tail coverts and tail, tawny, more or less marked with black.

Wilson's Snipe.
Jack Snipe.
Gallinago delicata.
See page 39.

Gallinago delicata.

Section 4. Toes, 4, without web; bill, over 1.10 and less than 1.90.

Tringa alpina pacifica.

Bill, decurved near tip; one or more of inner secondaries, almost entirely white; upper tail coverts, not white, barred with black; legs and feet, black. Spring birds have black on the belly, and back, rufous brown and black. Fall birds have the belly white and back gray.

Red-backed Sandpiper.
American Dunlin.
Tringa alpina pacifica.
See page 49.

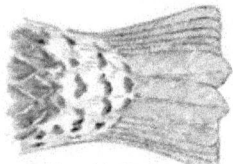

Tringa ferruginea.
Tail and upper tail coverts.

Bill, decurved near the tip; upper tail coverts, white, banded with black or dark brown.

Curlew Sandpiper.
Tringa ferruginea.
See page 50.

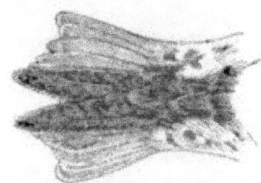

Tringa maculata.
Tail and upper tail coverts.

Tringa maculata.

Bill, nearly straight; back, marked with tawny and black; breast with numerous narrow, brown streaks; some of inner secondaries almost entirely white; lower rump and upper tail coverts, black; the feathers more or less tipped with buff. Two middle tail feathers longer than the others.

Pectoral Sandpiper.
Grass Bird.
Tringa maculata.
See page 45.

Bill, nearly straight; back, dark; feathers edged with ashy or buff; breast, grayish without brown streaks; one or more of inner secondaries almost entirely white; legs and feet, yellow in life, pale brown in dried skin.

Purple Sandpiper.
Tringa maritima.
See page 44.

KEY TO THE SHORE BIRDS OF NORTH AMERICA.

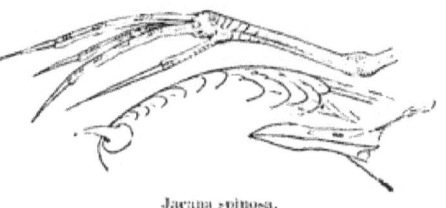

Tarsus, over 1.75; middle toe, over 1.60; primaries, pale yellowish green; **bend of wing with sharp spur.**

Mexican Jacana.
Jacana spinosa.
See page 85.

Jacana spinosa.

Section 5. Toes, 4, without web; bill, under 1.10.

Upper tail coverts, white; inner webs of primaries not speckled.

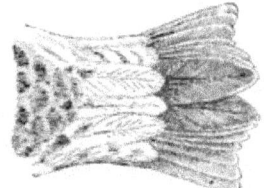

Tringa fuscicollis.

Tringa fuscicollis.

White Rumped Sandpiper.
Tringa fuscicollis.
See page 46.

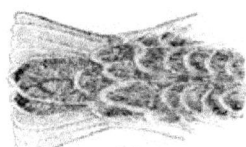

Middle upper tail coverts, smoky or dusky, often tipped with buff; inner webs of primaries not speckled; sides, white; **middle toe and claw, less than .95.**

Baird's Sandpiper.
Tringa Bairdii.
See page 47.

Tringa bairdii.
Tail and upper tail coverts.

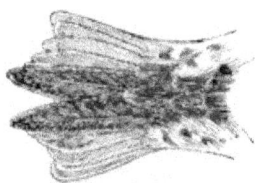

Middle upper tail coverts, black, often narrowly tipped with brownish buff; inner webs of primaries, not speckled; middle toe and claw, over .95; middle tail feathers decidedly longer than the rest.

Pectoral Sandpiper.
Grass Bird.
Tringa maculata.
See page 45.

Tringa maculata.
Tail and upper tail coverts.

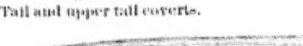

Inner webs of primaries speckled.

Buff-breasted Sandpiper.
Tryngites subruficollis.
See page 65.

Tryngites subruficollis.
First primary.

Section 6. Toes, 3; bill, over .60 in.

Bill, black; shoulder (lesser and middle wing coverts), brown; no web between toes; legs, black.

Sanderling Sandpiper.
Calidris arenaria.
See page 53.

Calidris arenaria.

Bill, thick; shoulder (lesser and middle wing coverts), ashy gray; legs, dull flesh color; a small web between toes.

Wilson's Plover.
Aegialitis Wilsonia.
See page 77.

Aegialitis Wilsonia.

Section 7. Toes, 3; bill, under .60.

Bill, orange at base, the tip, black; legs, dull flesh color; a black stripe from bill passing under eye.

**Semipalmated Plover,
Ring Neck.**
Aegialitis semipalmata.
See page 75.

Aegialitis semipalmata.

Bill, orange at base, the tip, black; legs, orange yellow; no black stripe from bill to eye; black breast band, not confluent; two middle tail feathers, **tipped with white.** Eastern species not found west of Rocky Mountains.

Piping Plover.
Aegialitis meloda.
See page 75.

Bill, orange at base, the tip, black; legs, orange yellow; no black stripe from bill to eye; **breast band, continuous** and not broken in the middle; two middle tail feathers, **tipped with white.** Eastern species not found west of Rocky Mountains.

Belted Piping Plover.
Aegialitis meloda circumcincta.

Bill, entirely black; legs, slate color; no black stripe from bill to eye; two middle tail feathers, **not tipped with white;** two outer tail feathers, white. Western species ranges from Texas and Kansas, west, to the Pacific Ocean. Accidental in Florida.

Snowy Plover.
Aegialitis nivosa.
See page 76.

Aegialitis nivosa.

GROUP IV.

Birds having wings from 5.50 to 6.75 inches long.

Section 1. Toes, 4, a small web between outer and middle toe; bill, over 1.80.

Macrorhamphus griseus.

Upper tail coverts and axillars, white, spotted or barred with dusky; bill, nearly straight.

Red-breasted Snipe, or Dowitcher.
Macrorhamphus griseus,
and
Western Red-breasted Snipe, or, Long-billed Dowitcher.
Macrorhamphus scolopaceus.
See page 41.

Section 2. Toes, 4, a small web between outer and middle toe; bill, under 1.80.

Totanus flavipes.

Tarsus and middle toe together, more than 2.60 inches long; outer primary, slate brown, without bars. Rump and upper tail coverts, white, more or less barred with brown; **legs, yellow.**

Summer Yellow-leg.
Totanus flavipes.
See page 58.

Tarsus and middle toe, together, less than 2.60; outer primary, slaty brown; rump and upper tail coverts, plumbeous gray; axillars, slaty gray, without bands. This species does not occur in Eastern United States.

Wandering Tattler.
Heteractitis incanus.
See page 61.

Tarsus and middle toe, together, less than 2.60; upper tail coverts, white or white barred with black; outer primary, slate brown, without bars.

Stilt Sandpiper.
Micropalama himantopus.
See page 42.

Outer primary, whitish, barred with dark brown.

Bartramian Sandpiper.
"Upland Plover."
Bartramia longicauda.
See page 64.

Section 3. Toes, 4, with small lobate webs.

Back, streaked tawny and black; underparts, more or less reddish brown.

                Red Phalarope.
                Crymophilus fulicarius.
                See page 32.

Section 4. Toes, 4, without webs; bill, over 2 in.

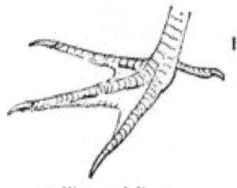

Lower belly, whitish or white; axillars, barred black and white.

                Wilson's Snipe.
                Jack Snipe.
                Gallinago delicata.
                See page 39.

Gallinago delicata.

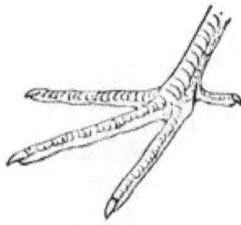

Belly, buff color; axillars, rufous brown.

                Woodcock.
                Philohela minor.
                See page 38.

Philohela minor.

Section 5. Toes, 4, without webs; bill, under 2.

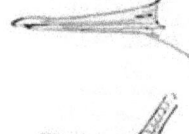

Rump, gray; upper tail coverts, whitish, banded or marked with black; inner webs of primaries not speckled.

                Knot.
                Tringa canutus.
                See page 43.

Tringa canutus.

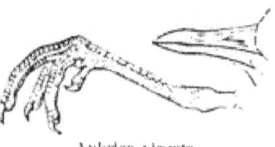
Aphriza virgata.

Rump, brownish or grayish; upper tail coverts, white, without black bands or spots; tail, dark, tipped with white; axillars, white. A Pacific species not found in Eastern North America.

Surf Bird.
Aphriza virgata.
See page 79.

Arenaria interpres.

Throat, white; rump and upper tail coverts, white (not banded); inner web of primaries not speckled.

Turnstone.
"Chicken Plover."
Arenaria interpres.
See page 80.

Throat, dusky; rump and upper tail coverts, white; head and neck, black; a white spot on lores and white streaks on forehead. A Pacific species not found in Eastern North America.

Black Turnstone.
Arenaria melanocephala.
See page 81.

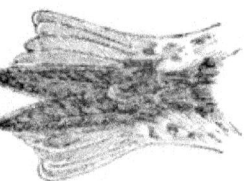
Tringa maculata.

Rump, blackish; middle upper tail coverts, black (not banded); inner web of primaries not speckled.

Pectoral Sandpiper.
Grass Bird.
Tringa maculata.
See page 45.

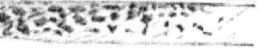
Tryngites subruficollis.

Inner webs of primaries, speckled.
Buff-breasted Sandpiper.
Tryngites subruficollis.
See page 65.

Section 6. Toes, 3.

Breast, with two black bands; under parts, white; rump and upper tail coverts, orange brown.

Killdeer Plover.
Ægialitis vocifera.
See page 74.

GROUP V.

Birds having wings from 6.75 to 9 inches long.

Section 1. Toes, 4, with more or less web; bill, curved upwards or straight; bill, over 2.60.

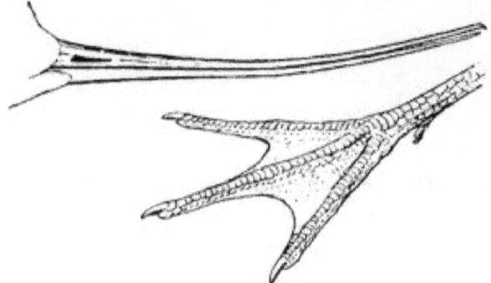

Axillars, white; belly, white; first primary, dark with dark shaft; top of head and nape, not black; bill, curved upward.

American avocet.
Recurvirostra americana.
See page 35.

Recurvirostra americana.

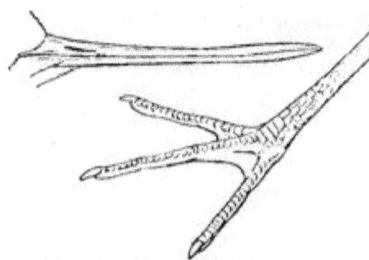

Axillars, white; belly, white; first primary, dark, with dark shaft; legs, very long, rose pink in life; bill, nearly straight; top of head and nape, black.

Black-necked Stilt.
Himantopus mexicanus.
See page 37.

Himantopus mexicanus.

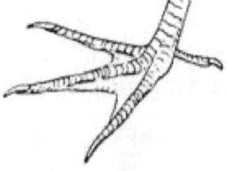

Axillars, smoky black; belly, white; terminal third of outer primary, black; the rest, white; bill, nearly straight.

Willet (*Symphemia semipalmata*) **and Western Willet** (*Symphemia semipalmata inornata*).

See page 60.

Symphemia semipalmata.

Axillars, dark gray, or sooty gray; belly, grayish white; **first primary, dark slaty brown with white shaft**; bill, curved upward; **upper tail coverts mostly white**.

Hudsonian Godwit.
Limosa hæmastica.
See page 56.

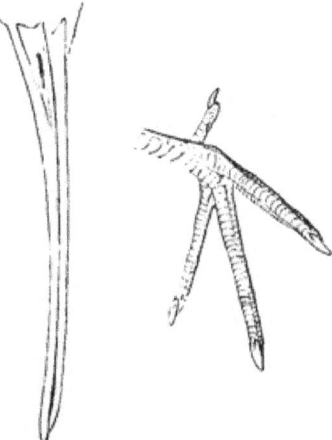

Axillars, rufous brown; upper tail coverts, not white; belly, buff; sometimes barred with dark brown; **primaries, pale rufous brown with numerous dark dots**; shaft of primaries, white; bill, curved upward.

Marbled Godwit.
Limosa fedoa.
See page 54.

Limosa fedoa.

Section 2. Toes, 4, without webs; bill, nearly straight; bill, over 2.60.

Axillars, banded with white and grayish brown; belly, pale brown, banded with dark brown; primaries, grayish brown; **outer webs, banded with pale brown or rufous brown**; shaft of primaries, dark; bill, nearly straight.

European Woodcock.
Scolopax rusticola.
See page 39.

Section 3. Toes, 4 (with small web); bill, slightly curved upward or straight; bill, under 2.60 and over 1.50.

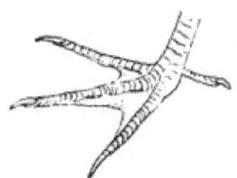

Axillars, smoky black; belly, white; outer primary terminal third, black, rest, white; rump, gray; upper tail coverts, white.

Willet.
Symphemia semipalmata.
See page 60.

Axillars, white, with few light brown dots near the ends; belly, white; outer primary dark, with shaft white; **rump, white without bars; upper tail coverts, white without bars; legs, olive green.**

Greenshank.
Totanus nebularius.
See page 58.

Axillars, dark gray; outer primary, brownish slate color; shaft, whitish; rump and upper tail coverts, **plumbeous gray.** Pacific species, does not occur in Eastern North America.

Wandering Tatler.
Heteractitis incanus.
See page 61.

Axillars, white, banded with brown; belly, white; outer primary, black; shaft, white; **rump, grayish brown; feathers, tipped with white; upper tail coverts, white, more or less barred with dark brown; legs, bright yellow.**

Winter Yellowlegs.
Greater Yellowlegs.
Totanus melanoleucus.
See page 57.

Axillars, pure white; belly, white; first primary, black with black shaft; **rump and upper tail coverts, black or brownish black;** legs, pinkish red in life.

Black-necked Stilt.
Himantopus mexicanus.
See page 37.

Section 4. Toes, 4, with small web; bill, nearly straight; bill, under 1.50.

Axillars, white, banded with brown; first primary, white, barred with dark brown; tip, dark.

Bartramian Sandpiper.
Upland Plover.
Bartramia longicauda.
See page 64.

Axillars, dark gray; without bars; first primary, dark (without bars); tail, not barred; back, plumbeous gray. Pacific species which does not occur in Eastern North America.

Wandering Tatler.
Heteractitis incanus.
See page 61.

Axillars, smoky black; first primary, dark (without bars); hind toe, very small; tail, barred dark brown and white; back, mottled.

Black-bellied Plover.
Squatarola squatarola.
See page 72.

KEY TO THE SHORE BIRDS OF NORTH AMERICA.   27

Section 5. Toes, 4, with small web; bill, curved downward; bill, over 2 inches long.

Axillars.

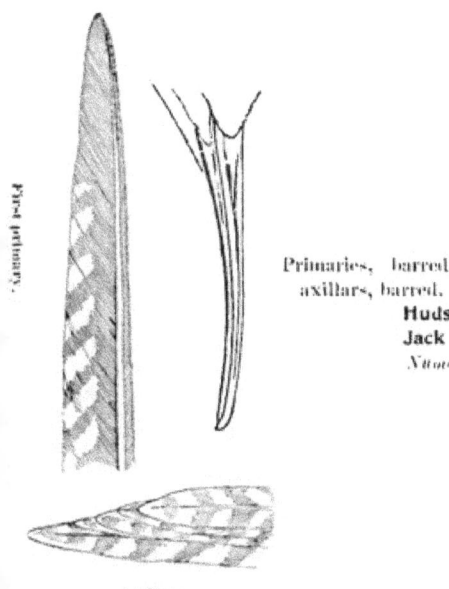

Primaries, barred;
axillars, barred.
Hudsonian Curlew.
Jack Curlew.
Numenius hudsonicus.

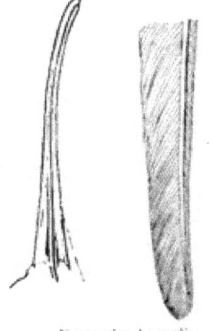
Numenius borealis.

Primaries, without bars;
axillars, barred.
Esquimaux Curlew.
Dough Bird.
Numenius borealis.

Section 6. Toes, 4 (cleft to base), without webs.

Upper tail coverts, white. Inhabits Pacific Coast of North America; not recorded from Eastern United States.

Surf Bird.
Aphriza virgata.
See page 79.

Section 7. Toes, 3.

Rump, orange brown; under parts, white with two black bands on the breast; axillars, pure white.

Killdeer Plover.
Egialitis vocifera.
See page 74.

Axillars, gray; rump, not orange brown.

Golden Plover.
Charadrius dominicus.
See page 73.

Charadrius dominicus.

Axillars, smoky black; rump and upper tail coverts, not orange brown.

This species has four toes (the hind toe being so small that it often escapes notice), and properly belongs to Group 5, Section 3, but owing to the fact that it is constantly looked for among the three-toed species it is included in both sections.

Black-bellied Plover.
Charadrius squatarola.
See page 72.

GROUP VI.

Birds having wings measuring from 9 to 12 inches long.

Section 1. Toes, 4; bill, curved downward.

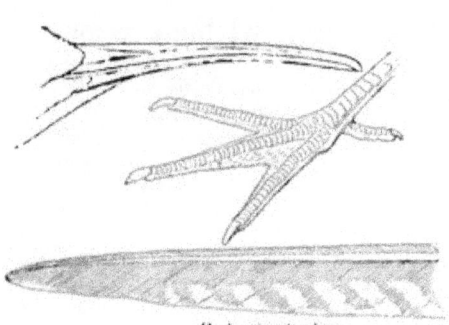

Hudsonian Curlew. Axillars.

Axillars, reddish brown with narrow black marks; belly, buff; bill usually over four inches.
    Long-billed Curlew.
    Sickle-bill Curlew.
Numenius longirostris.
See page 67.

Axillars, banded with slaty brown and dull white; belly, whitish; bill under four inches.
    Jack Curlew.
    Hudsonian Curlew.
See page 68.

Section 2. Toes, 4; bill, curved upward, or nearly straight.

Axillars rufous; primaries, rufous, dotted with black.

Marbled Godwit.
Limosa fedoa.
See page 54.

Limosa fedoa.

Section 3. Toes, 3; bill nearly straight.

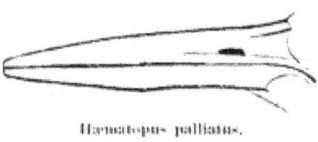

Hæmatopus palliatus.

Upper tail coverts, white; bill, red.
American Oyster-catcher.
Hæmatopus palliatus.
See page 82.

Upper tail coverts, mixed brown and white; bill, red. Pacific coast species does not occur in Eastern North America.
Frazar's Oyster-catcher.
Hæmatopus frazari.
See page 83.

General plumage and upper tail coverts, dark brownish black; bill, red. Pacific coast species does not occur in Eastern North America.
Black Oyster-catcher.
Hæmatopus bachmani.
See page 83.

ORDER LIMICOLÆ.

SHORE BIRDS.

FAMILY PHALAROPODIDÆ. PHALAROPES.

Wilson's Phalarope.   Red Phalarope.

The Phalaropodidæ or Phalaropes number but three species, all of which occur in North America. Although somewhat snipe-like in form, their webbed feet enable them to swim easily and gracefully, and they are often observed in flocks far out at sea. The female is larger and more highly colored than the male.

Genus CRYMOPHILUS Vieill.

CRYMOPHILUS FULICARIUS (*Linn.*).

Red Phalarope. Gray Phalarope.

SEA-GOOSE. WHALE-BIRD. BOWHEAD.

Crymophilus fulicarius.

Crymophilus fulicarius.

Adult male in summer: General under parts and sides of the neck, purplish brown; upper tail coverts, purplish brown, slightly darker than the under parts; crown of the head and base of the bill, smoky black; sides of the head, white, extending to the nape; rump, white; back, black; feathers, edged with tawny brown; primaries, dusky, the shafts being white, and the base of the feathers white; some of the secondaries, white; bill, yellowish brown, dark on the tip; feet, dull yellow.

Adult in winter: Head and under parts, white; a small patch around the eye and a nuchal crest, dusky; the under parts of the body are tinged with ashy gray on the sides; upper parts, grayish ash color; wings, showing a distinct white bar; bill, dusky; feet, brownish.

Length, 7.70; wing, 5.10; tail, 2.50; bill, .94; tarsus, .80.

The Red Phalarope is a maritime species which ranges from the far north, where it breeds, southward to the Middle States. On the Pacific coast it has been recorded as far south as Cape St. Lucas. During migrations it is not uncommon on some of our large interior lakes. It is gregarious in its habits, usually being observed in flocks far out at sea. It breeds in the far north. The eggs, which are four in number, are grayish or brownish, spotted with dark brown; they vary much in color, some eggs being described as greenish, heavily spotted with dull brown. The species is known to sailors by the name of "Whale Bird." Contrary to the usual rule among birds, the female is the handsomer and is also the larger.

PHALAROPUS LOBATUS (*Linn.*).

Northern Phalarope.

Adult female in summer: Upper plumage, dark plumbeous; the back, streaked with buff; sides and front of the neck, rufous brown; greater wing coverts, tipped with white; belly, white.

Adult male in summer: Similar to the female but paler; little, if any, rufous on the front of the neck.

Winter plumage: Upper plumage, grayish; forehead, sides of the neck, cheeks, and underparts, white; top of the head, dull gray; the feathers edged with dull white; a dusky spot on side of head and in front of the eye; breast, tinged with gray.

Phalaropus lobatus.

Length, 7.50; wing, 4.25; tarsus, .80; tail, 2; bill, .80 to .90.

The Northern Phalarope is another of our maritime species, at times not uncommon along our coast. It occurs on the Pacific coast as well as on the Atlantic side, and is common and breeds among the islands of Behring Sea. It is rather a more southern species than the preceding, and wanders in winter as far south as the Middle States, and to Mexico on the Pacific side. It occurs in the interior, being not uncommon on some of our large inland lakes. It swims easily and gracefully and is very much at home in the water on account of its webbed feet. The male of this species, as in the other Phalaropes, is smaller than the female and not so highly colored; he also takes upon himself many domestic duties usually assumed by the female; he sets on the nest, contrary to the usual custom, and devotes himself to the young chicks. The eggs are usually three or four, and are gray or grayish, blotched with chocolate brown. The nest, which is built on the ground, is usually composed of a little grass or moss.

PHALAROPUS TRICOLOR (*Vieill.*).

Wilson's Phalarope.

Phalaropus tricolor.

Adult female in summer: Crown and middle of back, pearl gray; nape, white; superciliary stripe, white; a dusky or black streak from the eye to the sides of the neck; sides of upper back, chestnut, bordering the gray; middle throat and breast tinged with pale rufous brown; chin, white; belly, white.
Adult male in summer: General resemblance to the female, but smaller and much paler in coloration; the crown and back, more brownish.
Adult in winter: Upper plumage, the feathers more or less edged with white; wings, fuscous, or gray brown; coverts, edged narrowly with white; under parts, white.
Female: Length, 9.10 to 10.05; wing, 5.20 to 5.40; tarsus, 1.25 to 1.40; bill, 1.25 to 1.40.
Male: Length, 8.10 to 9.10; wing, 4.60 to 4.90; tarsus, 1.20 to 1.30; bill, 1.20 to 1.30.

Wilson's Palarope is a more inland species than the preceding, and is by no means common on our coasts. It is abundant in some portions of the Mississippi Valley, and breeds in the United States from Illinois and Utah northward. The nest is simply a depression in the ground lined with a little grass, and the eggs are usually four in number, of dull brownish white, marked and spotted with dark chocolate brown. In winter the bird occurs in South America, and is claimed to wander as far south as Patagonia. In this, as with the other Phalaropes, the male assumes the duties of incubation.

Family RECURVIROSTRIDÆ. Avocets and Stilts.
Genus RECURVIROSTRA Linn.

This family consists of two or three genera, comprising ten or eleven species. While they usually obtain their food wading about in shallow water, they are web-footed and swim easily and gracefully.

RECURVIROSTRA AMERICANA *Gmel.*

American Avocet.

Adult in summer: Bill, very slender and curved upwards; feet, partly webbed; general plumage, white, becoming cinnamon brown on the head and neck, but remaining whitish at the base of the bill; primaries, black; most of the secondaries, white, forming a broad white patch on the wing; tail, ashy gray; legs, bluish; bill, black.

Adult in winter: Head and most of the neck, ashy gray; tail, ashy gray; rest as in summer plumage.

Length, 18 to 20; wing, 7.50 to 9; tail, 3.50; bill, 3.25 to 3.75; tarsus, 3.60.

This is a western species which ranges from the Great Slave Lake southward to Central America and the West Indies. It occurs commonly in Texas and along the Gulf Coast in winter, but is not common on the Atlantic coast north of Florida, although stragglers have been recorded from different points

extending as far north as the Bay of Fundy. It breeds from Texas and Illinois northward to the British Provinces. The eggs are from three to four in number, pale grayish olive spotted with very dark brown.

Genus **HIMANTOPUS** Briss.

HIMANTOPUS MEXICANUS (*Mull.*).

Black-necked Stilt.

Adult: Top and sides of the head and back of the neck and back, black; under eyelid and a spot above and behind the eye, forehead, and sides of the head under the eye, white; rest of under parts, with the rump and upper tail coverts, white; tail, ash gray or pearl gray; bill, black; legs, red in life.

The immature bird has the upper parts brownish; feathers, edged with whitish; wings, blackish, some of the feathers tipped with white; wing coverts, edged with buff or tawny brown; under parts, white mottled with black and tawny brown.

Length, 15.50; wing, 9; tail, 3; tarsus, 4.25; bill, 2.70.

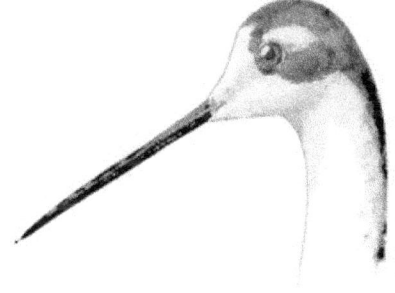

Himantopus Mexicanus.

This species is common throughout tropical America, occasionally being found in the Mississippi Valley as far north as Minnesota. It is accidental on the Atlantic coast north of Florida and Southern Georgia, although it has been recorded as far north as Long Island and Maine. It is common in Florida after April 1, and breeds from Florida and the Gulf States northward up the Mississippi Valley to Minnesota, and west to Utah and Oregon. The eggs are laid on the ground, sometimes on a little grass, and again without any nest whatever. They are three or four in number and pale brown or dull white, spotted and marked with chocolate brown.

395909

Family SCOLOPACIDÆ. Snipes, Sandpipers, etc.

A very large family, numbering some 100 species, about 45 of which occur in North America.

Genus PHILOHELA Gray.

PHILOHELA MINOR (*Gmel.*).

American Woodcock.

Adult: Bill, long, corrugated near the end; upper parts variegated with tawny rufous, brown and black; head, rufous brown, banded on the occiput with alternate bands of black and tawny brown; a line of black from the eye to the bill; a narrow patch on the sides of the lower head, blackish; throat, pale rufous or rufous white; under parts of the body, pale rufous; quills, brownish; tail feathers, dusky, tipped with ash; bill, pale brown, yellowish at the base; legs, reddish; bill, light brown, paler and yellowish at base.

Length, 11.50; wing, 5.45; tarsus, 1.30; bill, 2.90 to 3.05.

Common on the Atlantic coast, from the British Provinces to the Gulf of Mexico, west to Kansas and Dakota, breeding throughout its range, but more commonly north of the Carolinas. It is occasionally observed in Florida in winter, and is very abundant in the swamps of Alabama and Louisiana at some seasons. The eggs are usually four, mottled and spotted with brown.

The European Woodcock (*Scolopax rusticola*) is larger than our bird, and may be easily distinguished from it by the heavily barred under parts, and having the wings also barred with rufous brown. The outer primary is not emarginate, as in the American species. There are numerous records of the occurrence of this bird in eastern North America.

Genus GALLINAGO Leach.

GALLINAGO DELICATA (*Ord*).

Wilson's Snipe.

English Snipe.   Jack Snipe.   Gutter Snipe.

Gallinago delicata.

Adult: Bill, long and straight, being slightly enlarged near the tip and showing numerous small pits; *no web* between outer and middle toes; general upper parts, dark brown, tawny brown, pale yellow, dull white; top of the head, black with middle stripe of tawny brown; outer web of first primary, white; greater wing coverts, brownish or dusky with white tips; axillars and under surface of wings, barred with black; upper tail coverts, tawny brown barred with black; tail feathers, chestnut brown with subterminal black bar; tips, white, and feathers marked with black at the base; under parts, white; breast and throat, speckled and lined with brown; sides of the body, brownish barred with dull black.

Length, 11; wing, 5; tail, 2.20; tarsus, 1.25; bill, 2.40 to 2.60.

The English Snipe, or Wilson's Snipe, is one of our best known game birds and is very abundant in suitable localities during the migrations. It ranges from Canada and British Columbia, south, in winter to the West Indies, and even to South America. It breeds from the latitude of New England northward. The nest is placed on the ground, and the eggs are three or four, usually of a grayish ash color blotched with chocolate brown, heaviest at the largest end. They measure 1.50 x 1.15. In some localities in the Southern States, during the winter months, thousands of these birds are killed on the marshes where they collect on some especially good feeding ground. When first disturbed they utter a peculiar *cheep* as they rise from the ground, often repeated during their flight, which is very irregular, making them one of the most difficult birds to shoot.

The European Snipe (*Gallinago gallinago*) has not as yet been taken in the United States; but it has been recorded from Greenland and the Bermuda

English Snipe, or Wilson's Snipe.

Islands. It somewhat resembles our bird, but the tail feathers usually number fourteen, although this character is not constant. It has no standing as a North American species, except from its occurrence in Greenland.

Genus MACRORHAMPHUS Leach.

MACRORHAMPHUS GRISEUS (*Gmel.*).

Dowitcher. Red-breasted Snipe. Brown Back.

Adult in summer: A small web between the outer and middle toes; upper parts, blackish;

Macrorhamphus griseus.

feathers edged with tawny brown; top of head, blackish, mottled with tawny brown; under parts, red brown, banded on the sides; throat and breast only slightly spotted, entirely without spots in some plumages; **tail and upper tail coverts, barred with black**; bill, dark olive.

Adult in winter: Upper parts, grayish, the feathers showing faint edges of buff on the back; chin and superciliary stripe dull grayish white; breast, gray, showing slight traces of tawny (often entirely absent); rest of under parts, dirty white, mottled on the crissum; tail, banded dark brown or black and white.

Length, 10.30; wing, 5.70; tarsus, 1.30; bill, 2.20 to 2.50.

The Red-breasted Snipe, or Dowitcher, ranges from the Arctic Circle, where it breeds, to South America, being common in the United States during the migrations. It is a well-known bird to sportsmen, and its long, snipe-like bill will always distinguish it from other species of shore birds, except the Wilson's Snipe, from which it may be known at a glance by the small web between the outer and middle toes, and by its differently marked tail and tail coverts. It occurs in flocks, and where it has not been persecuted by gunners, is very tame and unsuspicious, and comes readily to decoys. The eggs are dull buff or pale olive speckled with dark brown.

MACRORHAMPHUS SCOLOPACEUS *Say*.

Long-billed Dowitcher. Western Red-breasted Snipe.

Resembles the preceding species, but the bill is longer, and in breeding plumage, the sides of the body are more distinctly barred with black; the throat and breast more rufous, and the general plumage more highly colored.

In winter the principal difference is one of size.

Length, 11; wing, 6.05; tarsus, 1.55; bill, 2.20 to 2.60.

This species, which is closely allied to the eastern bird, in breeding plumage, may be distinguished from it by its longer bill, barred sides, and richer coloration. In winter they differ but little except in length of bill. It ranges from Alaska to the Mississippi Valley, and south to Mexico, occasionally wandering to the Atlantic coast, occurring somewhat regularly in winter in some of our southeastern States. The eggs are not distinguishable from those of the preceding species.

Genus MICROPALAMA Baird.

MICROPALAMA HIMANTOPUS (*Bonap.*).

Stilt Sandpiper.

Micropalama himantopus.

Adult in summer: Legs, **dull olive green;** entire plumage, dull white and dark brown, being banded on the under parts with dark brown, and the upper parts streaked with dark brown; a stripe of chestnut brown above and below the eye, the former extending backwards and joining at the occiput; secondaries edged with tawny brown.

Adult in winter: Upper plumage, gray; feathers on the back, edged with white; breast, pale buff, slightly mottled, shading into dull white on the upper throat; belly, dull white; bill, black.

Length, 8.20; wing, 5; **tarsus, 1.65;** bill, 1.60.

This species ranges from the Arctic Circle to South America, being not uncommon on the Atlantic coast at times during the migrations; abundant in Florida, in March, on the marshes along the east coast. It has somewhat the appearance of a very diminutive summer yellow-legs, but its legs are **olive green** instead of yellow.

It breeds in the far north; the eggs are usually 4, pale, buffy white or grayish white, spotted with brown, and measure 1.42 x 1.

Genus TRINGA Linn.

Subgenus TRINGA.

TRINGA CANUTUS (*Linn.*).

Knot. Robin Snipe.

REDBREAST SANDPIPER. BLUE PLOVER. GRAYBACK.

Tringa Canutus.

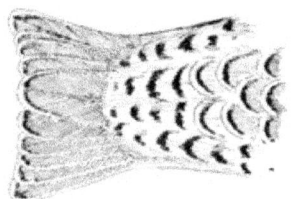

Upper tail coverts.

Summer plumage: Top of the head, buffy white, mixed with blackish. Upper plumage, variegated, with black, white, and buff. Upper tail coverts, barred with black and white, often marked with buff; entire under parts, except the lower abdomen, rufous brown; lower abdomen, white; axillars, white, with long, arrow-shaped markings of dark gray.

Winter plumage: Entire upper plumage, ashy gray, the feathers having pale, brownish shafts, showing indistinct fine lines; upper tail coverts, white, banded with brown; under parts, white; the breast and sides with indistinct, irregular dots and bands of brownish gray; belly, and under tail coverts, pure white; axillars, white, with irregular, arrow-shaped, brownish gray marking.

Length, 10.60; wing, 6.70; tarsus, 1.25; bill, 1.30.

The Knot is a cosmopolitan species which breeds in the Arctic regions, and is common in the United States during migrations. It winters from Florida to South America, a few remaining in Florida all winter. The eggs are known only from a single specimen taken near Fort Conger, by General Greeley, and described as "light pea green, closely spotted with brown, in small specks about the size of a pinhead." (See Merriam, Auk, II, 1885, p. 313.)

Subgenus ARQUATELLA Baird.

TRINGA MARITIMA *Brünn..*

Purple Sandpiper.

Tringa maritima.

Summer plumage: Top of head, dark gray; back, blackish, the feathers edged with grayish white; rump and upper tail coverts, black or brownish black; a small black spot in front of the eye; breast, grayish, having the appearance of being spotted with black, which black spotting also appears on the sides of the body. This is caused by the feathers of the breast being dark at the base tipped with white, and on the sides of the body the feathers are white with black spots near the tips; axillars, white.

Winter plumage: Similar, but lacking the black spots on the breast and sides, which is replaced by pale gray; bill, yellowish at base.

Length, 9.25; wing, 5; tarsus, .92; bill, 1.30.

The Purple Sandpiper is a northern species, breeding in high latitudes, but occurring in winter on the Atlantic coast to the Middle States, and occasionally to Florida. Some birds remain on the New England coast all winter, frequenting rocky ledges. It breeds from Northern Hudson's Bay to Greenland. The eggs, usually four, are buff, sometimes tinged with olive and mottled with brown, and measure 1.40 x 1.05. It does not occur on the Pacific coast.

Subgenus ACTODROMAS Kaup.

TRINGA MACULATA *Vieill.*

Pectoral Sandpiper.

GRASS BIRD.

Tringa maculata.

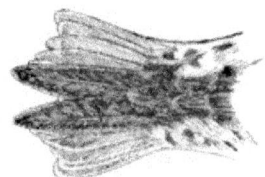

Tringa maculata.

Adult in summer: Head and upper parts, dark brown, the feathers edged with tawny; throat and belly, white; breast, pale brown, the feathers narrowly streaked with dark brown or blackish; upper tail coverts, black; two middle tail feathers longer than the others; basal half of bill, dull greenish yellow.
Winter plumage: Breast, brownish buff, and upper parts, more brownish.
Length, 9; wing, 5.25; tarsus, 1.05; bill, 1.15.

This species ranges from the Arctic regions to South America, being common on the Atlantic coast during migrations. It breeds in the far north. The eggs are four, greenish buff, mottled with brown, heaviest at the larger end, and measure 1.50 x 1.09 (Murdoch). The Pectoral Sandpiper may always be distinguished from the next species by its blackish middle upper tail coverts.

TRINGA FUSCICOLLIS *Vieill.*

White-rumped Sandpiper.

Tringa fuscicollis.

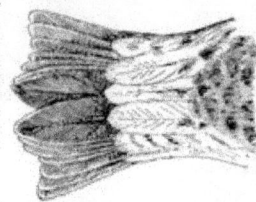

Rump and upper tail coverts.

Adult in summer: Top of the head and back, mottled with black, dull white, and buff; throat, white; breast, finely lined with dark brown; belly, white; an imperfect superciliary line of dull white; rump, dusky gray, and most of upper tail coverts, white.

Adult in winter: Top of the head and back, grayish, some of the feathers marked with dark brown, giving it a slightly mottled appearance on the back and top of the head; upper throat, white; breast, ashy gray, the shafts of the feathers showing brown; belly, white; forehead, whitish, extending in an imperfect superciliary line.

Length, 6.85; wing, 4.95; tarsus, .95; bill, .95.

This species may readily be distinguished by its white upper tail coverts. It ranges from the Arctic regions south to Central and South America and the West Indies, being one of our common beach birds during the migrations. It breeds in high latitudes. The eggs, which are usually four, are buff or olive, spotted and dotted with dark brown, and measure about 1.27 x .94 (Ridgway).

TRINGA BAIRDII (Coues).

Baird's Sandpiper.

Tringa bairdii.

Summer plumage: Crown, dark brown, mixed with buff; nape and upper back, narrowly streaked with buff and dark brown; the feathers of the back, dark brown, narrowly edged with white; **upper tail coverts, blackish; the feathers usually narrowly tipped with buff;** two central tail feathers, very slightly longer than the others, — not decidedly longer, as in Tringa maculata; chin, white; breast, brownish buff, showing very faint and narrow streaks of brown; under parts, buffy white, with a faint tinge of buff, sometimes entirely white.

Winter plumage: Similar, but paler.
Immature: Resembles the adult, but has the feathers of the back and wing coverts tipped with white.
Length, 7.40; wing, 4.50 to 4.90; tarsus, 1; bill, 1.

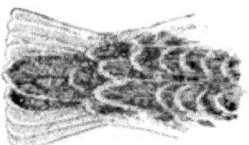

Tringa bairdii.

Baird's Sandpiper is an inland species, which ranges from the British Provinces to Western South America. It is a rare bird on the Atlantic coast, although occasionally taken during the migrations. It may always be distinguished from the White-rumped Sandpiper by its dark upper tail coverts, and from the Pectoral Sandpiper by the absence of the elongated middle tail feathers. This species is not known to occur on the Pacific coast. It breeds in the Arctic regions. The eggs are usually four, pale brown or cream buff, spotted with chestnut brown.

TRINGA MINUTILLA (*Vieill.*).

Least Sandpiper.

Adult in summer: Very small; upper parts marked with tawny black and pale buff; under parts, whitish on the throat, shading into ashy gray, showing faint brownish lines on the breast; rest of the under parts, white; bill, dull black. No web between toes.

Tringa minutilla.   Tringa minutilla.

Adult in winter: Similar to the above, but grayer in color.
Length, 5.50 to 6.50; wing, 3.60; tarsus, .75; bill, .85.

The Least Sandpiper may be distinguished from the Semipalmated Sandpiper by the absence of the small web between the toes, so noticeable in that species.

It is common throughout North and South America, breeding north of the United States. It is very abundant on our shores during the migrations, and is one of the species known to gunners by the name of Peep. The eggs are three to four, generally pale buff, mottled with chestnut brown.

Subgenus PELIDNA Cuv.
TRINGA ALPINA PACIFICA (Coues).
Red-backed Sandpiper.

Adult in summer: Bill slightly bent down near the tip; the feathers on the top of the head and back, bright rufous brown, heavily blotched with black on the central part of each feather; throat, whitish; breast streaked with black; belly, black on the upper part; lower belly and crissum, white; bill, black.

Tringa alpina pacifica. *Adult in winter:* Upper parts, gray, slightly mottled; throat, white; breast, gray, the shafts of the feathers dark brown; belly and under tail coverts, pure white; a faint, dull white superciliary line.

Length, 8; wing, 4.75; tarsus, 1.10; bill, 1.65.

The Red-backed Sandpiper, American Dunlin, or Leadback, as it is sometimes called, ranges throughout North America from the Atlantic to the Pacific, breeding far northward. It is common on the coasts during migrations, and is naturally tame and unsuspicious. It frequents both beaches and

Summer. Winter.

marshes, and is usually observed in flocks. It has been found breeding along the shores of Behring Sea and on Melville Peninsula. The eggs are usually four, bluish white or buffy white, marked with chocolate brown, mostly at the larger end. They measure about 1.40 x 1. *The peculiar decurved bill will always distinguish this species.*

The European Dunlin, *Tringa alpina* (Linn.), which is very closely allied to the American form, is claimed to occur in Eastern North America. It differs chiefly in being somewhat paler in coloration, and averages slightly smaller, but these differences are apparently by no means constant.

TRINGA FERRUGINEA *Brunn.*

Curlew Sandpiper.

Tringa ferruginea.

Adult, summer plumage: Head, neck, and under parts, except crissum, cinnamon brown, or chestnut brown; upper parts, blackish, mixed with reddish brown; upper tail coverts, whitish, marked with black; bill, slightly decurved.

Adult, winter plumage: Upper plumage, gray or brownish gray, the feathers showing dark shaft streaks, a superciliary stripe, and under parts, white; the chest faintly tinged with gray; rump, white; upper tail coverts, white, marked with black.

Young birds have the upper plumage dusky, the feathers edged with buff; breast tinged with buff.

Length, 7.25 to 9; wing, 4.75 to 5.20; bill, 1.40 to 1.60; tarsus, 1.08 to 1.20.

The Curlew Sandpiper is an old world species, which, although it has been taken a number of times on our shores, can only be considered as a rare straggler. Specimens have been recorded from Canada, Nova Scotia, Massachusetts (3) and Long Island. It has also been taken in Alaska.

Subgenus ANCYLOCHEILUS Kaup.

Genus EREUNETES Illiger.

EREUNETES PUSILLUS (*Linn.*).

Semipalmated Sandpiper "Peep."

Ereunetes pusillus.

Adult in summer: Heavily marked with dark brown and tawny on the top of the head and back; forehead, whitish; under parts and throat, white; breast, mottled with grayish; belly, white; *toes with small web*.

Adult in winter: Upper parts, grayish, mottled with brown on the head, and the feathers showing dark brown, edged with white on the back; throat, white; breast, very faintly washed with gray, which is sometimes very indistinct; belly and under tail coverts, white; *toes with small web*.

Length, 6; wing, 3.75; tarsus, .75; bill, .80.

Ranges from the Arctic Sea, southward, to the West Indies and South America. It is common on the beaches and marshes of the Atlantic coast during migrations, where it is called "Peep" by gunners. In size and general appearance it somewhat resembles the Least Sandpiper (another

"Peep" of the gunning fraternity), but may always be distinguished from that species by the semi-palmated or partly webbed toes. It breeds in the arctic regions. The eggs are buff white, mottled with brown and chocolate brown, heaviest on the larger end. They are usually four in number and measure about 1.20 x .85.

EREUNETES OCCIDENTALIS *Laur.*

Western Semipalmated Sandpiper. "Peep."

In summer plumage this species may be distinguished from *E. pusillus* by its longer bill (which is oftentimes 1.25 inches in length, while *E. pusillus* rarely, if ever, has the bill one inch), by its decidedly streaked breast, and the feathers of the back being margined with rufous. The winter plumages, however, differ but little, if any, and occasionally specimens of the Western Sandpiper occur which have the bill less than one inch, and in such cases it is almost impossible to distinguish them from the winter examples of *E. pusillus;* therefore it is perhaps as well for the sportsmen to consider all of those birds having bills less than one inch in length to be *E. pusillus*, and if it is important to have the identification absolutely accurate, they could easily obtain the opinion of some professional ornithologist to settle such a fine point.

Length, 6; wing, 3.85; tarsus, .82; bill, .90 to 1.15.

The Sandpiper occasionally occurs on the Atlantic coast, although its home is properly in the West. It ranges from the arctic regions and Alaska, where it breeds, south to the Gulf of Mexico and South America. It is common in Florida in winter, more so on the west than on the east coast.

The eggs are four, pale buff, dotted with dark brown or chocolate brown.

Genus CALIDRIS Cuv.

CALIDRIS ARENARIA (*Linn.*).

Sanderling. Bull-peep.

Calidris arenaria.

Adult in summer: Head, throat, and upper parts, except rump, reddish brown, black, and white; throat and breast, pale rufous brown, with more or less dark spots; belly and under parts, pure white; bill and feet, black; toes, *three*.
Adult in winter: Top of the head and back, ash gray, the shafts of the feathers being brown; forehead and entire under parts, white; bill, black.
Length, 8.10; wing, 5.05; tarsus, 1; bill, 1.

This cosmopolitan species may always be distinguished by the absence of the fourth toe, being the only sandpiper* occurring on our coast having three toes. It is common during migrations, ranging from the Arctic

---

* Plovers also have three toes.

Calidris arenaria.

regions to the West Indies and South America, being numerous on the Atlantic and Pacific, as well as in the interior. It breeds in the far North.

The eggs, which are usually four, are light olive and buff, spotted with brown.

Genus LIMOSA Brisson.

LIMOSA FEDOA (*Linn.*).

Marbled Godwit. Red Marlin.

Adult in summer: Bill, curved slightly upward; upper parts, mottled with black, and tawny; upper throat, white, rest of throat finely streaked with brown; breast, pale tawny, the feathers banded irregularly with brown; belly, tawny, sometimes without bands; bill, dull flesh color in its basal half, the rest blackish; inner webs of outer primaries, speckled with black; tail, barred with black; axillars, irregularly banded with dark slaty brown, in some instances merely showing an indication of bands but always with more or less irregular marks or dots where the bands are not perfect.

Adult in winter: Top of the head, brown, streaked with pale brown; feathers of the back, dark brown, edged with tawny; chin, white; throat, pale buff, faintly barred with brown; inner webs of outer primaries, speckled with black.

Length, 19; wing, 8.80; tarsus, 2.80; bill, 3.50 to 4.50.

The Marbled Godwit, or Marlin, occurs throughout North America, breeding in the interior, from the Missouri region northward. On the Pacific coast it ranges from Alaska to Central America, and is very common in some localities in Lower California and Texas. It is not very common anywhere on the Atlantic coast, although it occurs at times in some numbers in Florida.

Limosa fedoa.

The eggs are three or four, dull brownish ash color, blotched and mottled with gray and grayish brown.

The Pacific Godwit, *Limosa lapponica baueri* (Naum.), has once been recorded from La Paz, Lower California (Bryant). It occurs in Alaska in summer, but its true habitat is Australia and the Pacific Islands.

LIMOSA HÆMASTICA (*Linn.*).

Hudsonian Godwit.

Limosa hæmastica.

Summer plumage: Upper parts, dark brown, the feathers showing spots of pale rufous brown on the edges; **rump, white**; tail, black, tipped with whitish; chin, whitish with pale rufous; rest of under parts, dark rufous brown; the feathers of the throat more or less streaked with black, and the feathers of the breast and belly, faintly edged with black; axillars, very dark slate color, almost black.

Winter plumage: Upper plumage, dull gray; feathers of the back more or less edged with dark brown; chin, whitish; breast, pale gray; shafts of the feathers on the sides of the breast, brownish; belly, grayish buff, sometimes buffy white; under tail coverts, whitish; axillars, dark, smoky gray.

Length, 15; wing, 8 to 8.60; tarsus, 2.40; bill, 2.80 to 3.40.

Limosa hæmastica.

The Hudsonian Godwit, Ring-tailed Marlin, or Goosebird, as it is often called, ranges from the arctic regions to Southern South America. It is common in suitable localities during migrations east of the Rocky Mountains but does not occur on the Pacific coast. This fine bird, like many other of our birds, is gradually becoming less and less numerous. It breeds in the far North. The eggs, which are generally four in number, are olive brown, spotted with dark brown, and measure about 2.20 x 1.35.

TOTANUS MELANOLEUCUS (Gmel.).

Greater Yellow-legs. Winter Yellow-legs.

Totanus melanoleucus.

Adult in summer: Bill, nearly straight; upper plumage, mottled with white and black; upper tail coverts, white, barred with black; throat, white, streaked with black; chin, whitish; breast, white, heavily streaked with black; rest of under parts, white, irregularly marked with black; middle of belly, nearly always pure white; bill, black, and **legs, yellow**.

Adult in winter: Top of the head and neck, streaked white and dark brown; back, brown, the feathers narrowly edged with whitish; chin, white; breast, white, narrowly lined with dark brown; rest of under parts, white; bill, black, and **legs, yellow**.

Length, 14; wing, 7.75; tarsus, 2.45; bill, 2.30.

The Winter Yellow-leg, and his near relative, the Summer Yellow-leg, are probably the best known representatives of any of our shore birds. The clear, sharp, whistling note, repeated rapidly four or five times, descending in semi-tones down the scale, is known to gunners throughout the land. Although not as abundant as in former years, it is still numerous in many localities during the migrations. Its range extends from the sub-arctic regions south to Chili and Buenos Ayres, breeding from Illinois northward.

The eggs are usually three or four, buff brown, a pale buff, spotted with dark brown.

TOTANUS FLAVIPES (Gmel.).

Yellow-legs. Summer Yellow-legs.

Totanus flavipes.

Upper parts, grayish brown, mottled with whitish on the back; chin, white; throat, white, streaked with pale grayish brown; breast, mottled; belly, white; tail feathers, banded brown and white; upper tail coverts, white, more or less barred with black; bill, black; **legs, yellow**. Resembles the Greater Yellow-legs, but is smaller. Axillars, white, barred with brown.
Length, 10.80; wing, 6.45; tarsus, 2.10; bill, 1.45.

Ranges from the sub-arctic regions southward to South America. The Summer Yellow-legs resembles the preceding species, but is smaller, and is usually much more tame and unsuspicious than the Winter Yellow-legs, and comes to decoys more readily. It is very common on the Atlantic coast during the migrations. The note is similar, though perhaps not so sharp and loud. It breeds in the interior, from Minnesota and Illinois northward. The eggs are pale buff, dotted and blotched with dark brown and chocolate brown.

The Green-shank, *Totanus nebularius* (Gmm.), a European species, was taken by Audubon, near Cape Sable, Florida, in 1832. It somewhat resembles the Yellow-leg, but has the rump and lower back, white, and the legs, greenish.

Subgenus RHYACOPHILUS Kaup

TOTANUS SOLITARIUS (Wils.).

Solitary Sandpiper.

Totanus solitarius.

Adult in summer: Top of head and back, and upper tail coverts, bronzy green, dotted with white; under parts, white; the breast, thickly streaked and dotted brown; bill, greenish brown (in life), dusky, terminally; **axillars, white, heavily barred with smoky black.**

Adult in winter: Upper parts, including upper tail coverts, olive brown, showing a faint, greenish gloss when held in the light, the feathers faintly dotted with dull white; throat, white; breast, streaked with brown; rest of under parts, white; axillars, heavily barred.

Length, 8.50; wing, 5.30; tarsus, 1.20; bill, 1.30.

Axillars.

This species ranges from the sub-arctic regions, southward, to South America. It occurs with us during the migrations, and breeds from Pennsylvania and Illinois northward. It is usually observed about inland ponds and rivers, rarely frequenting the salt marshes. The eggs, which are described from a single specimen taken by Jenness Richardson, in Vermont, and described by Dr. Brewer, are light drab with small rounded brown markings, having a few faint purplish marks on the larger end.

The European Green Sandpiper, *T. ochropus* (Linn.), has been recorded from Nova Scotia. It somewhat resembles *T. solitarius*, but has the upper tail coverts white.

Genus SYMPHEMIA Raf.

SYMPHEMIA SEMIPALMATA (*Gmel.*).
Willet.
Humility. Stone Curlew.

Symphemia semipalmata.

Adult in summer: Upper plumage, gray, streaked on the head with dark brown, and the central portion of many of the feathers has the back blotched with brown; chin, white; throat, white, dotted with brown; under parts, dull white, the feathers on the sides barred with brown and washed with tawny; **axillars, black**; bill, bluish, dusky toward end.

Adult in winter: Bill, nearly straight; upper plumage, gray, showing faint indications of whitish on the tips of some of the feathers on the back; chin, whitish; throat, ashy gray; the rest of under parts, white, showing ashy on the sides of the body; **axillars, smoky black**. The broad, white band on the wings is a distinguishing character of this bird when flying, formed by the basal portion of the primaries and some of the secondaries being white.

Length, 15.50; wing, 8.10; tarsus, 2.30; bill, 2 to 2.30.

Symphemia semipalmata.

Ranges throughout temperate North America, south to the West Indies and South America. It breeds from New England to Florida, although it does not breed commonly north of the Carolinas. It is a very common species in Florida, where it frequents the beaches and marshes, where its broad, white wing band and peculiar whistle will always distinguish it from other species.

The eggs are three in number, sometimes four, deposited on the ground with scarcely any indication of a nest. They are pale buff in color, spotted with chocolate brown, heaviest on the larger end.

SYMPHEMIA SEMIPALMATA INORNATA *Brewst.*

Western Willet.

General resemblance to the preceding species but the upper parts are paler and not so heavily marked with black; the breast shows usually more buff color. In winter plumage it resembles the eastern form except in size, the bill usually being longer and the bird somewhat larger.

Length, 15.60; wing, 8.50; tarsus, 2.55; bill, 2.30 to 2.80.

Common throughout Western North America extending east to Mississippi Valley and Gulf States, being numerous in Florida in winter. It breeds from Texas to Manitoba.

HETERACTITIS INCANUS (*Gmel.*).

Wandering Tattler.

Heteractitis Incanus.

A faint superciliary stripe of white and loral stripe of black; upper plumage, slaty gray, showing a greenish gloss when held in the light; throat, whitish; breast, ashy gray, indistinctly barred with dull brown; belly, white; axillars, gray.

Length, 10.60; wing, 7; bill, 1.50; tarsus, 1.30.

The Wandering Tattler does not occur in Eastern North America, but it ranges from Norton Sound, Alaska, to the Galapagos Islands, and also occurs in the Hawaiian Islands and Kamchatka. It does not seem to be very abundant anywhere, and we know but little regarding its habits. It probably breeds in the far North but the nest and eggs have never as yet been taken.

PAVONCELLA PUGNAX (*Linn.*).

European Ruff.

Adult male (spring plumage): Feathers of the head and neck, elongated, forming a wide collar or ruff, generally white and buff; rest of upper parts variegated with buff, black, dull white, and ochre; primaries, dark brown, with white shafts; under parts, white, heavily marked with blackish brown on the breast and sides; face covered with yellowish tubercles; bill, dark brown, lighter at the base; iris, dark brown; legs, brownish yellow.

Length, 12.50; wing, 7; tail, 2.70; tarsus, 2; bill, 1.55.

Adult female (Reeve): No ruff, as in the male; head and neck, sandy brown, mottled faintly with dark brown; upper parts variegated with black and brown, and tinge of reddish; under parts, dull white, mottled on the breast and sides with brown. The rest as in the male.

Male (winter plumage): Lacking the ruff and tubercles on the face; plumage resembles the female. The specimens figured in the plate represent two adult males and a female. The variation of the coloration of the plumes of the former is very great; the ruff may be black, white, or chestnut barred, and banded in various ways, or plain white, but generally showing a tinge of buff or black.

The European Ruff has been taken a number of times in Eastern North America. Specimens have been recorded from Long Island, Massachusetts, Maine, Ohio, and elsewhere. (Coues, Key to N. A. Bds., p. 641.)

European Ruff.

Genus BARTRAMIA Less.

BARTRAMIA LONGICAUDA (*Bechst.*)

Bartram's Sandpiper. Upland Plover. Field Plover.

Summer plumage: Upper plumage, dark brown, or brownish black; the feathers edged with pale buff; upper throat, white; lower throat, pale buff lined with dark brown; breast, pale buff with arrow-shaped markings of dark brown; belly, pale buff; axillars banded with dark slaty brown and white; outer primaries, white, banded with brown on the inner webs.

Winter plumage: Similar but paler.

Length, 11.75; wing, 6.60; tarsus, 1.30; bill, 1.20.

Bartramia longicauda.

The Bartram's Sandpiper, or Upland Plover, ranges in North America, from Alaska to Nova Scotia, and south to South America, but mainly east of the Rocky mountains. It is not very common anywhere on the Atlantic coast nowadays, although at one time it was an abundant species on our hills during the migrations. A good diagnostic character of this species is the barred primary. It breeds from Virginia northward. The eggs are four or five in number of a pale brownish color, mottled with chocolate brown near the larger end.

Genus TRYNGITES Cabanis.

TRYNGITES SUBRUFICOLLIS (*Vieill.*).

Buff-breasted Sandpiper.

Tryngites subruficollis.

Summer plumage: Upper plumage, buff, mottled with black; the feathers on the back, black, edged with buff; under parts, having a mottled buff and white appearance, caused by the exposed portion of the feathers being buff, narrowly tipped with white; the feathers are dark slate color at base, but the slate color is entirely concealed; **inner web of first primary, white, speckled with dark brown;** the inner primaries and secondaries, narrowly tipped with white, showing a sub-terminal band of black; axillars, white.

Winter plumage: Similar, but paler.

Length, 8; wing, 5.30; tarsus, 1.25; bill, .85.

Not abundant on the Atlantic coast, usually keeping to the interior. It breeds from British America northward to the Arctic Ocean. It ranges, in winter, south to South America.

First primary.

The eggs are described as three or four, pale, buff white, spotted and lined with dark brown and purplish brown.

Genus ACTITIS Illiger.

ACTITIS MACULARIA (*Linn.*).

Spotted Sandpiper.

Adult in summer: Top of the head and back, olive green, showing bronzy reflections when held in the light, some of the feathers on the back irregularly marked with brown; under parts, white, marked with large, round black spots.

Summer.   Actitis macularia.   Winter.

Adult in winter: Above, olive, showing a faint, bronzy luster when held in the light; feathers on the back, faintly tipped with dusky; wing coverts, narrowly banded with tawny; throat, white; breast, faintly tinged with ashy; under parts, including belly and under tail coverts, white; mandible and edge of the maxilla, pale wax yellow (in life), rest of bill, black.

Length, 8; wing, 4.20; tarsus, 1; bill, 1.

The Spotted Sandpiper ranges throughout North America to Northern South America. It is a very common bird on our coast, being usually seen alone or in pairs. It is common on small streams of fresh water and also on sand beaches, usually selecting rocky places. It breeds nearly throughout its range, the eggs being buff white or pale brown, spotted with dark brown, mostly on the larger end.

Genus **NUMENIUS** Brisson.

NUMENIUS LONGIROSTRIS *Wils.*

Long-billed Curlew. Sickle Bill.

Adult in summer: Upper parts marked with buff and black; tail feathers, alternately banded with tawny buff and brown; throat, whitish; rest of under parts, pale reddish brown, becoming very light on the belly; breast, narrowly striped with brown on the middle of the feathers; bill, very long; **axillars, rufous brown.**

Long-billed Curlew.   Esquimaux Curlew.   Hudsonian Curlew.

Adult in winter: General plumage, tawny brown; the back, blackish, mottled with buff; top of the head, dark brown; the feathers, edged with tawny; throat, white; under parts, pale buff brown; feathers on the lower throat and upper breast, finely lined with dark brown; bill, very long and curved downward; bill, black, becoming dull lilac brown on basal half of the mandible; **axillars, rufous brown.**

Length, 26; wing, 10.50; tarsus, 2.30; bill, very variable, measuring from 2.50 to 9.

The immature of this species has the bill nearly straight, but quite short, sometimes not exceeding two inches in length.

The Long-billed Curlew ranges from temperate North America south to Central America and the West Indies. It breeds in the Southern Atlantic

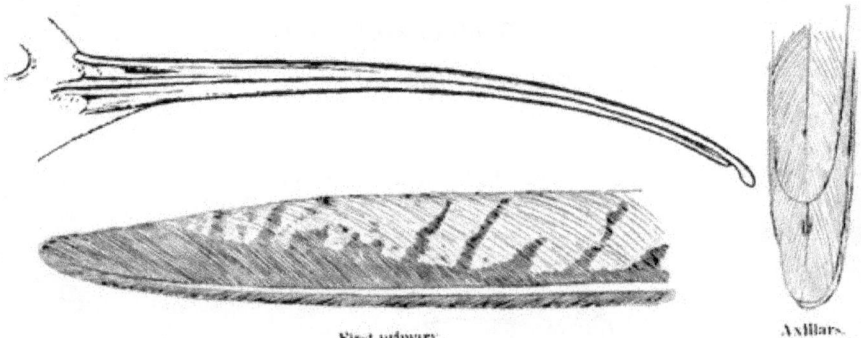

First primary. Axillars.

States, but is now not common on the Atlantic coast, and is becoming less so every year. The eggs are described as olive gray in color, spotted with chocolate brown.

NUMENIUS HUDSONICUS *Lath.*

Hudsonian Curlew. Jack Curlew.

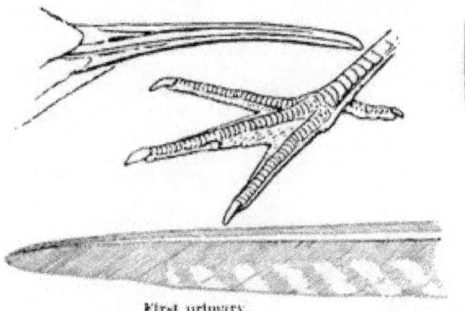

First primary. Axillar.

Adult in summer: Bill curved downward; top of the head, brown, with a stripe of buffy white through the middle; rest of upper parts, dark brown; feathers, pale on the edges, giving a slightly mottled appearance to the back; chin, white; throat and breast streaked and faintly barred with brown and white; belly, dull white; an imperfect superciliary stripe; **axillars, pale buff, barred with slaty brown; first primary, barred on inner web.**

Adult in winter: Similar to the last, but showing much more tawny mottled on the upper parts, and the under parts, paler; a superciliary stripe of dull white; **axillars, pale buff, barred with slaty brown; first primary, barred on inner web.**

Length, 17.50; wing, 9.20; tarsus, 2.30; bill, 3 to 3.75.

Ranges throughout North America, breeding in the arctic regions and migrating in the fall to the Gulf of Mexico and South America. During the migrations it is a well known bird to gunners on the Atlantic coast, but is not easy to kill as it pays but little attention to decoys and will not be enticed within shot, however good the imitation of its trilling whistle. It breeds in

the far North. The eggs are usually three or four, grayish olive spotted with brown.

This species may always be distinguished from the Esquimo Curlew by the barred inner web of the primary and the buffy stripe on the middle of the crown, and from the Long-billed Curlew by the heavily barred axillars.

NUMENIUS BOREALIS (*Forst.*).

Eskimo Curlew. Dough Bird.

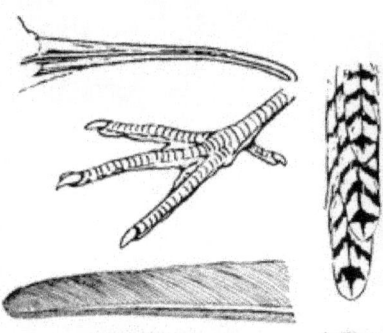

Adult in summer: Bill, curved downward, general plumage above, mottled black and tawny; chin, whitish; throat and under parts, tawny buff, marked on the throat with dark brown, and on the breast with arrow-like brown markings; flank and sides of the body, tawny, the arrow-like marks being much heavier and larger; top of the head showing no central stripe of buffy white; **inner web of first primary without bars; axillars, barred with slaty brown.**

Adult in winter: Lacking the tawny color of the summer plumage; more whitish on the under parts, otherwise the markings being similar.

Length, 13; wing, 8.10; tarsus, 2; bill, 2.75 to 3.50.

It ranges from the arctic regions, where it breeds, southward to South America. It is more common in the interior than on the coast, although it was at one time a common bird in the New England States during the migrations. It prefers the fields to the beaches, being often found in company with the Golden Plover. The eggs are described as pale olive gray, spotted with dark brown, mostly at the larger end.

Family CHARADRIIDÆ. Plovers.

The Plovers are a cosmopolitan family, numbering something less than one hundred species, fifteen of which occur in North America, including exotic

Black-bellied Plover (Winter).   Piping Plover.   Semipalmated Plover.

stragglers. As a rule they have but three toes, although two genera, Squatarola and Vanellus, have four. The tarsus is reticulate and the toes are partly webbed.

Genus CHARADRIUS Linn.
Subgenus SQUATAROLA (Cuv.).
CHARADRIUS SQUATAROLA (Linn.).
Black-bellied Plover. Beetle Head. Black-breast.

Charadrius squatarola.

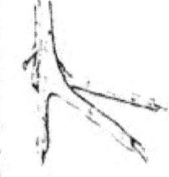

Adult male in summer: Upper parts, smoky black; feathers, edged with dull white; breast, black in highest plumage, but usually showing traces of whitish on the feathers; abdomen and under tail coverts, white; forehead, white; **axillars, smoky black.**

Adult in winter: Upper plumage, brownish, mottled with gray; under parts, white; breast and sides, streaked with ashy brown; bill, black.

Length, 11.25; wing, 7.45; tarsus, 1.85; bill, 1.10.

Winter.

Summer.

This species may always be distinguished from the Golden Plover by the presence of a rudimentary hind toe and the black axillars, which are gray in the Golden Plover.

It ranges from the arctic regions to South America, being common on the Atlantic coast during migrations. It breeds in the far North; the eggs are usually four, pale olive brown, spotted and speckled with brown and black. The note is a soft whistle, si-sol, of the scale.

Subgenus **CHARADRIUS** Linn.

CHARADRIUS DOMINICUS *Mull.*

American Golden Plover. Green Plover.

Summer plumage: Back and upper parts, smoky black; feathers marked and edged with yellow and brown; sides of the breast, whitish; rest of under parts, throat, and sides of the head, including eye, black; forehead, whitish, a white stripe extending backwards over the eye; down the sides of the neck; bill, black; **axillars, gray.**

Winter plumage: Upper parts, brownish, feathers marked with dull tawny yellow or buff; under parts are dull white, streaked with ashy brown or gray on the breast and sides.

Length, 10.25; wing, 7; tarsus, 1.55; bill, .92.

Summer.   Winter.

This species may always be distinguished in any plumage from the Black-breast Plover by the gray axillars and the absence of the small rudimentary hind toe.

It ranges from the arctic regions where it breeds to South America, being common on the coast during migration. It was at one time abundant in New England in the early fall, but its numbers have decreased greatly within the past few years. While it prefers the hills and fields, it also frequents the marshes and occasionally the sand flats. The eggs are buff colored or brownish white, mottled and marked with chocolate brown.

The European Lapwing, *Vanellus vanellus* (Linn.), has once been recorded from Long Island (Dutcher, Auk III., 1886, p. 438).

Genus AEGIALITIS Boie.

Subgenus OXYECHUS Reich.

AEGIALITIS VOCIFERA (*Linn.*).

Killdeer Plover.

Adult: Crown and back, brownish gray, feathers tipped with rufous; a ring around the neck, lores, and a patch on the breast, black; forehead, white; throat and spot behind the eye and a band around the neck, white; lower breast and belly, white; tail coverts and rump, bright rufous; tail, rufous and gray, tipped with black and white; bill, black.

Length, 10.65; wing, 6.55; tarsus, 1.40; bill, .75.

Aegialitis vocifera.

The Killdeer Plover ranges on the Atlantic coast from Newfoundland to the West Indies and South America, although as a rule it is a rare bird in New England. In November, 1888, a great flight of these birds occurred along the coast of Massachusetts, and hundreds of them were offered for sale in the markets. The writer killed several near Hyannis, Mass., at that time. It is abundant in the Gulf States in winter, and its shrill notes are very familiar sounds to the Southern sportsmen.

It breeds throughout its range. The eggs are described as pale brownish white, mottled and lined with chocolate brown.

Subgenus AEGIALITIS Boie.

ÆGIALITIS SEMIPALMATA Bonap.

Semipalmated Plover. Ring-neck.

Ægialitis semipalmata.

Adult in summer: Forehead, white, succeeded by a band of black; rest of head, back, and wings, ashy gray; a dull stripe of blackish passing from the bill under the eye to the sides of the neck; a small patch of white back of the eye; a broad band of black on the breast; rest of under parts, white; throat, white, extending around the neck and joining at the back in a very narrow stripe; secondaries, tipped with white; bill black, orange at the base; a bright orange ring around the eye; toes webbed at base.

Adult in winter: Similar, but the black marking replaced by brownish or gray.

Length, 6.80; wing, 4.75; tarsus, .93; bill, .50.

Ranges from the arctic regions to South America, breeding from Labrador northward. It is a common species on our coast during the migrations. The eggs are usually four, pale brownish white, spotted with chocolate brown.

The Little Ring Plover, *Ægialitis dubia*, an Asiatic species, is of accidental occurrence on the coast of California.

ÆGIALITIS MELODA (Ord.).

Piping Plover.

Ægialitis meloda.

Adult in summer: Upper parts, ashy; forehead, white, separated from the ash color of the head by a narrow patch of black; a band on the side of the breast, black; under parts and a ring around the neck, white.

Adult in winter: Similar to the summer plumage, but the black marking replaced by grayish buff, or ashy gray, instead of black; bill, orange at base, tipped with black.

Length, 7.20; wing, 4.80; tarsus, .90; bill, .50.

Ranges from Labrador to the West Indies and South America. It is a common species on the New England coast during the migrations, and is a regular winter visitant to Florida, although not abundant.

The Piping Plover breeds from Virginia to Labrador and Newfoundland. The eggs are usually four, dull white, marked with dark brown.

ÆGIALITIS MELODA CIRCUMCINCTA Ridgw.

Belted Piping Plover.

It is similar to *A. meloda*, but differs in having an unbroken black band from the sides joined in front of the neck forming a complete collar. This form occurs on the coast during the migrations. It breeds commonly in the interior from Northern Illinois north to Winnipeg, and migrates in winter to the Gulf of Mexico.

ÆGIALITIS NIVOSA Cass.

Snowy Plover.

Ægialitis nivosa.

Adult in summer: Upper plumage, light brownish gray, or buffy gray; forehead and superciliary region, lores, and under parts, pure white; a patch of black on the front part of the crown, bordering the white forehead; a small patch of black on the ear coverts, and a black patch on each side of the chest; bill, entirely black.

The female is similar, but has the black marking duller, faint grayish.

Winter plumage, male: General resemblance to the adult male in summer, but the black marking replaced by grayish brown, or buff gray; legs, black; bill, entirely black, which is a good diagnostic character.

Length, 6.30 to 7.10; wing, 4.20 to 4.35; bill, .60; tarsus, .88 to 1.05.

The Snowy Plover is common on the Pacific coast, from Northern California to Central and South America. It also occurs, in winter, in many of the Gulf States, and is not uncommon in northwestern Florida.

The eggs are laid in a mere depression in the sand. They are usually four in number, buff gray color, spotted with black.

Subgenus OCHTHODROMUS Reich.
ÆGIALITIS WILSONIA (*Ord*).

Wilson's Plover.

Ægialitis Wilsonia.

Adult male: Above, ashy brown; forehead, white, extending into a faint superciliary stripe of dull black on the crown; throat, white, continuing on the sides of the neck, nearly joining upon the nape; a black, pectoral band, the feathers edged with white, becoming brown upon the sides; under parts, white; two central tail feathers, brown, the others showing increasing markings of white to the outer tail feathers, which are white; bill, black (large and stout); legs, pinkish.

Female and immature birds have the pectoral band brown, and no black on the head.

Length, 7.45; wing, 4.60; tail, 1.90; tarsus, 1.16; bill, .90.

Wilson's Plover occurs both on the Atlantic and Pacific coasts of North America, ranging from Long Island and Lower California southward to the West Indies and South America. Stragglers have been recorded as far north as Nova Scotia.

It frequents the sandy beaches and flats often in company with other species. It breeds from Virginia southward, the nest being a mere depression in the sand. The eggs are usually three in number, dull white, spotted and marked with chocolate brown.

Subgenus PODASOCYS Cours.

ÆGIALITIS MONTANA (*Towns.*).

Mountain Plover.

Ægialitis montana.

Summer plumage: Forehead and stripe over the eye, white; a stripe of black from the bill to the eye; front of crown, black; rest of upper plumage, including back and crown, grayish buff tinged with tawny; throat, white; breast, faintly washed with tawny buff; rest of under parts, white; axillars, white.

Winter plumage: Entire upper plumage, including crown, buff gray; feathers, faintly edged with tawny; throat, whitish; breast, more or less tinged with buff or brownish white; rest of under parts, pure white; axillars, white.

Length, 8.75; wing, 6; tarsus, 1.60; bill, .95.

"Temperate North America from the great plains, westward; accidental in Florida." (A. O. U.)

The Mountain Plover is strictly an inland species, rarely, if ever, frequenting the shores and marshes of the coast. It ranges from Central Kansas to the Rocky Mountains, migrating in winter, southward, to Southern and Lower California and Mexico, and probably along the Gulf coast in some localities, as it has been taken in Florida. It breeds from Kansas northward to the British boundary. The eggs are three, grayish olive, spotted with dark brown.

Family APHRIZIDÆ Surf Birds and Turnstones.
Subfamily APHRIZINAE. Surf Birds.
Genus APHRIZA Audubon.

APHRIZA VIRGATA (*Gmel.*).

Surf Bird.

Aphriza virgata.

Aphriza virgata.

Adult, summer plumage: Upper plumage, streaked and mottled with black and grayish white; the scapulary plumes and some of the back feathers, rufous with a sub-terminal band of black tipped with white; rump, dark brownish slate color; upper tail coverts, white; tail feathers, black, narrowly tipped with white; under parts, dotted with black on the throat and spotted with black on the breast and sides of the body, the breast showing more or less gray.

Winter plumage: Entire upper parts, dark brownish gray; throat, whitish; breast, brownish gray; rest of under parts, white, dotted more or less with brownish gray; rump, dark gray; upper tail coverts, white; tail, brownish black tipped with white; axillars, white.

The Surf Bird occurs on the Pacific coast of North and South America, from Alaska to Chili, but is apparently nowhere common. The nest and eggs are unknown. It has never been taken in Eastern North America.

Subfamily ARENARIINÆ. Turnstones.

Genus ARENARIA Brisson.

ARENARIA INTERPRES (*Linn.*).

Summer.　　　Arenaria Interpres.　　　Winter.

Turnstone.

Chicken Plover. Brant Bird. Calico Back.

Adult in summer: General upper parts, mottled and variegated with black, white, rufous and tawny; throat and breast, black and white; rest of under parts, white; tail, with subterminal band of black, tipped with white.

Adult in winter: Above, light, streaked and dashed with dark brown; an imperfect band of dark brown on the jugulum; chin and upper part of the throat, white; sides of breast, like the back; rest of the under parts, white; a distinct white band on the wing; rump, white, but with a broad patch of black on the upper tail coverts; tail, dark brown, the tips and basal half of the inner feathers, and nearly two thirds of the outer feathers, white; legs, reddish orange; bill, black.

Length, 8.65; wing, 5.70; tail, 2.60; tarsus, 1; bill, .95.

The Turnstone, Chicken Plover, or Brant Bird, as it is variously called, is a cosmopolitan species. It is common on both coasts of North America, and occasionally occurs in the interior, on the shores of the larger lakes and rivers. It breeds in the arctic regions. The eggs, which are usually four, are dull clay color, marked with pale brown or grayish brown, and measure about 1.55 x 1.15.

It is common on the Atlantic coast during migrations, frequenting sandy beaches and flats, sometimes in small flocks, but often alone or with a single companion.

It has received its name from the manner in which it turns over small pebbles and shells while searching for the small crustaceans and insects upon which it feeds.

ARENARIA MELANOCEPHALA (Vig.).

Black Turnstone.

Arenaria melanocephala.

Adult, summer plumage: General plumage, brownish black, with a tinge of orange; forehead and breast, streaked with white; a white spot on the lores, and a white bar on the wing; belly and crissum, white.

Adult, winter plumage: Similar, but lacks the white on the head and breast.

Length, 9; wing, 5.90; bill, 1; tarsus, 1.

The Black Turnstone occurs only on the Pacific coast of North America, ranging from Point Barrow, Alaska, to Lower California. It breeds from Alaska to British Columbia. The eggs are three or four, dull clay color, marked with dusky brown.

Family HÆMATOPODIDÆ. Oyster-catchers.

Genus HAEMATOPUS Linn.

HÆMATOPUS PALLIATUS Temm.

American Oyster-catcher.

Hæmatopus palliatus.

Winter plumage, male: Head and neck, blackish or very dark brown; back, brown; lower part of breast and rest of under parts, white; eyelids, rump, tips of wing coverts, part of secondaries, and basal portion of the tail feathers, white; bill, reddish orange, darkening at the tip (in summer, deep red); legs, flesh color.

Length, 17.40; wing, 10.05; tail, 4.35; tarsus, 2.30; bill, 3.50.

The American Oyster-catcher is a strictly maritime species, frequenting the beaches and flats exposed by the tide, where it searches for clams and small bivalves. It is not uncommon in suitable localities on the Atlantic coast from New Jersey southward, and on the Pacific side from Lower California to Patagonia. Stragglers have been taken on the coast of Maine and Massachusetts. The eggs are laid in a depression in the sand. They are usually three or four, buff white in color, blotched and spotted with chocolate brown, and measure about 2.20 x 1.50.

HÆMATOPUS FRAZARI. *Brewster.*

Frazar's Oyster-catcher.

Geographical distribution: "Lower California (both coasts), north, to Los Coronados Islands." (A. O. U.)

Three specimens of this interesting species were procured by Mr. Frazar north of La Paz, on the Gulf of California. It is described as "differing from the North American bird in having a stouter and more depressed bill, little or no white on the eyelids; the back, scapulars, and wing-coverts, richer and deeper brown, the primaries and tail feathers, darker; the upper tail-coverts, more or less varied with brown and white; the lateral under tail-coverts, marked with brown; the bend of the wing and greater under primary coverts, mottled with black and white; from the Galapagos species in the rather shorter bill and distinctly brown (instead of sooty black) back, scapulars and wing-coverts, dark markings on the under tail-coverts, and the greater amount of white on the under primary coverts; from both the above-mentioned species in the broad zone of mottled black and white feathers extending across the breast. Extreme measurements, three specimens, all males: wing, 9.75 to 10.27; tail, 3.90 to 4.26; tarsus, 2.18 to 2.30; bill, length from nostril, 2.35 to 2.37; from feathers, 2.99 to 3.05; depth at angle, 49.53." (Brewster, Auk. V., Jan., 1888, p. 84.)

HÆMATOPUS BACHMANI *Aud.*

Black Oyster-catcher.

Adult. Sp. char.: Head and neck, black; rest of plumage, dark brown or brownish black; bill, vermilion red; legs and feet, flesh color.
Length, 16.50; wing, 9.20; bill, 2.70; tarsus, 1.70.

Geographical distribution: "Pacific coast of North America, from the Aleutian Islands to La Paz, Lower California." (A. O. U.)

Jacanas.

Family JACANIDÆ.

JACANA SPINOSA (*Linn.*).

Mexican Jacana.

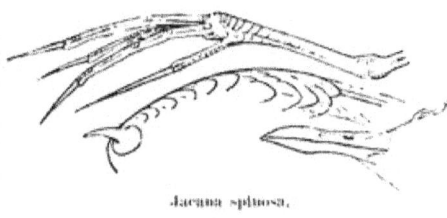

Jacana spinosa.

Sp. char. Adult male: Bill and comb, pale orange; bare skin at the base of the lower mandible, pale bluish white; head, neck, and upper breast, dark, lustrous green; back and wing coverts, purple, shading into rich golden brown near the rump; rump and tail coverts, purple; under parts, dark purple, showing a tinge of dark rufous on the crissum; most of the primaries and secondaries, yellow, edged with brown; tail, rufous brown; carpal spur (a horny spur on bend of wing), pale orange; legs and feet, dull olive.

Length, 9; wing, 5; tail, 2.25; tarsus, 2.20; bill, 1.30.

The immature plumage is very different, but the bird in any plumage can hardly be confounded with any other North American species.

Geographical distribution: "Lower Rio Grande Valley, Texas, south to Panama, Cuba and Hayti." (A. O. U.)

The Jacana is given a place among our birds from its casual occurrence in the Lower Rio Grande Valley. It is a fresh water species, frequenting the ponds and creeks.

INDEX.

A.

	PAGE
Actitis maculuria	66
Actodromas	45
Ægialitis	74
meloda	75
" circumcincta	76
montana	78
nivosa	76
semipalmata	75
vocifera	74
wilsonia	77
American Avocet	35
Dunlin	48
Oyster Catcher	82
Woodcock	38
Aphriza	79
virgata	79
Arenaria interpres	80
melanocephala	81
Arquatella	44
Avocet	35

B.

Baird's Sandpiper	47
Bartramia longicauda	64
Bartram's Sandpiper	64
Beetle Head	72
Belted Piping Plover	76
Black-bellied Plover	72
Black-breast Plover	72
Black Turnstone	81
Black Oyster Catcher	83
Black-necked Stilt	37
Brant-bird	80
Brown-back	41
Buff-breasted Sandpiper	65
Bull Peep	53

C.

Calico-back	80
Calidris arenaria	33

Charadriidæ

	PAGE
Charadriidæ	71
Charadrius	72
dominicus	73
squatarola	72
Chicken Plover	80
Crymophilus fulicarius	32
Curlew Sandpiper	50
Jack	68
Hudsonian	68
Long billed	67
sickle bill	67
Eskimo	70

D.

Doe Bird	70
Dough Bird	70
Dowitcher	41
Dunlin	49

E.

English Snipe	39, 40
Ereunetes	51
occidentalis	52
pusillus	51
Eskimo Curlew	70
European Dunlin	49
European Ruff	62
European Woodcock	39

F.

Field Plover	64
Frazar's Oyster Catcher	83

G.

Gallinago delicata	39
Godwit, Hudsonian	56
Godwit, marbled	54
Golden Plover	73
Goosebird or Godwit	54
Grass Bird	45

INDEX.

	PAGE
Gray Phalarope	32
Greater Yellow-legs	57
Green Plover	73
Green Shank	58
Gutter Snipe	39

H.

Hæmatopodidæ	82
Hæmatopus palliatus	82
Heteractitis incanus	61
Himantopus mexicanus	37
Hudsonian Curlew	68
Hudsonian Godwit	56
Humility, or Godwit	56
Humility	60

J.

Jacana	85
spinosa	84, 85
Jack Curlew	68
Jack Snipe	39

K.

Killdee	74
Killdeer Plover	74
Knot	43

L.

Least Sandpiper	48
Limicolæ	31
Limosa	54
fedoa	54
hæmastica	56
lapponica baueri	55
Long-billed Curlew	67
Long-billed Dowitcher	41

M.

Macrorhamphus	41
griseus	41
scolopaceus	41
Marbled Godwit	54
Micropalama himantopus	42
Mountain Plover	78
Marlin	54

N.

Northern Phalarope	33
Numenius	67
borealis	70
hudsonicus	68
longirostris	67

O.

	PAGE
Oyster Catcher	82, 83

P.

Pavoncella pugnax	62
Pectoral Sandpiper	45
Peep	51, 52
Pelidna	49
Phalarope	31
Phalarope, Red	32
" Gray	32
" Northern	33
" Wilson's	34
Phalaropidæ	31
lobatus	33
tricolor	34
Philohela minor	38
Purple Sandpiper	44
Plover, Green	73
Golden	73
Belted Piping	76
Black-bellied	72
Black-breasted	72
Field	64
Killdeer	75
Mountain	78
Piping	75
Ring-neck	75
Snowy	76
Semipalmated	75
Upland	64
Wilson's	77

R.

Recurvirostra americana	35
Red Marlin	54
Red Phalarope	32
Red-breasted Snipe	41
Red-backed Sandpiper	49
Rhyacophilus	59
Ring-neck	75
Robin Snipe	43
Ruff	62

S.

Sanderling	50
Sandpiper, Bartramian or Bartram's	64
Baird's	47
Buff-breasted	65
Curlew	50
Least	48
Pectoral	45
Purple	44
Red-backed	49
Red-breasted	43
Semipalmated	51

INDEX.

	PAGE
Sandpiper,—*Continued*.	
Solitary	39
Spotted	66
Stilt	42
Western semipalmated	51
White-rumped	46
Scolopacidae	38
Scolopax rusticola	39
Semipalmated Sandpiper	51
Sickle-bill Curlew	67
Squatarola	72
Snipe	38, 39
Snowy Plover	76
Solitary Sandpiper	39
Spotted Sandpiper	66
Stilt	35
Stilt Sandpiper	42
Stone Curlew	69
Summer Yellow-legs	58
Surf Bird	79
Symphemia semipalmata	60
" " inornata	64

T.

Totanus	57
ochropus	59
flavipes	58
melanoleucus	57
nebularius	58
solitarius	59
Tringa	43
alpina	49

Tringa—*Continued*.	
alpina pacifica	
bairdii	
canutus	
ferruginea	
fuscicollis	
maculata	
maritima	
minutilla	
Tryngites subruficollis	
Turnstone	

U.

Upland Plover	

W.

Wandering Tattler	
Western Red-breasted Snipe	
Semipalmated Sandpiper	
Willet	
White-rumped Sandpiper	
Wilson's Snipe	
Plover	
Phalarope	
Willet	
Winter Yellow-legs	
Woodcock	

Y.

Yellow-legs	

www.ingramcontent.com/pod-product-compliance
Lightning Source LLC
Chambersburg PA
CBHW020258090426
42735CB00009B/1137